INTERNATIONAL CENTRE FOR MECHANICAL SCIENCES

COURSES AND LECTURES - No. 100

ANTONI K. OPPENHEIM
AND
MOSTAFA M. KAMEL
UNIVERSITY OF CALIFORNIA

LASER CINEMATOGRAPHY OF EXPLOSIONS

LECTURES DELIVERED DURING THE COURSE
ON EXPERIMENTAL METHODS IN MECHANICS
OCTOBER 1971

UDINE 1971

SPRINGER-VERLAG WIEN GMBH

Originally published by Springer-Verlag Wien New York in 1972

ISBN 978-3-211-81179-5   ISBN 978-3-7091-2860-2 (eBook)
DOI 10.1007/978-3-7091-2860-2

PREFACE

Explosions are familiar events of every-day's life be they of cosmic, terrestrial, or man made origin. Their effects range from the cosmic displays of star explosions and solar flares, to the destructiveness of lightnings, earthquakes, volcanic eruptions, and meteoroid impacts; from the potential hazards of air pollution and sonic boom, to the more immediate danger of thermonuclear holocaust. Thus the idea of harnessing, or controlling, these destructive processes for useful and productive purposes is most attractive. A prelude to controllability, however, is a through understanding of the phenomena under consideration. The scientific endeavor directed towards this aim has become known as the Gasdynamics of Explosions. The description of its fundamental features can be found in the text of Oppenheim (*). Shortly then, the objective of this novel branch of fluid mechanics is to contribute to the knowledge of the interrelationships between the exothermic rate processes that are at the heart of explosions and the non-steady gasdynamic phenomena that manifest, in effect, their mechanical work output, the energy that can be employed for destructive or productive purposes.

(*) Oppenheim, A.K., Introduction to Gasdynamics of Explosions, International Center for Mechanical Sciences (CISM), Udine, Italy, 1971.

From its inception, the pace of theo-
retical progress in the field of Gasdynamics of Ex-
plosions has been geared to, as well as controlled by,
advances in the experimental techniques utilized in
observing the explosion phenomena. Since the signi-
ficant period of the generation of an explosion has a
duration, in most cases, of less than a microsecond,
the event should be observed with an instrument having
a nanosecond resolution in time. In order to visualize
the magnitude of this resolving power, one has only
to realize that the ratio of one nanosecond to a sec-
ond is approximately equivalent to one second in the
life span of a thirty year old man.

This experimental capability has only
recently become available, primarily by the control-
lable laser beam associated with solid state electron-
ic techniques. An example of the latter is the fast
response pressure transducer as described in the texts
of Oppenheim (*) and Soloukhin (**).

The objective of this monograph is to
present the technique developed by the exploitation
of the controllable laser beam: the laser cinemato-
graphy. For this purpose, described here first is the
apparatus and some of the results obtained by its use
in the study of spherical blast waves in explosive

(**) Soloukhin, R.I., <u>Udarnye Volny i Detonatsia v Gazakh</u>
(Shock Waves and Detonation in Gases), 175 pp., Gos. Izd. Fiz.
Mat. Literatury, Moscow, 1963; (Transl. by B.W. Kavshinoff,
Mono Book Corp. Baltimore, 1966).

gas mixtures. This is followed by the description of the optical methods that can be employed in this connection to record the refractive index fields associated with explosion phenomena. The electronic features of the apparatus are then described in detail. Finally, included here also is a description of a continuous flow device for safe and accurate mixing of explosive gases under laboratory conditions.

Similarly as the INTRODUCTION TO GAS-DYNAMICS OF EXPLOSIONS (*), the subject matter of this monograph is based entirely on the results of a research program which was conducted over a number of years at the University of California in Berkeley. These results could not have been obtained without the contribution of many collaborators. Essential assistance in this connection has been provided by Drs. G.J. Hecht and P.A. Urtiew with the help of Mr. K. Hom who was primarily responsible for all the technical work associated with the construction and maintenance of the experimental apparatus. Special thanks are due to Professor F.J. Weinberg for the valuable help he gave us in the development of optical techniques, to Dr. J.W. Meyer who became most skilled in the operation of the experimental equipment, and to Mrs. S. Frolich for the care and attention she gave in the production of the manuscript. The authors wish to take this opportunity to express their thanks to their wifes and children who provided most of the inspiration for their work.

The authors would like to acknowledge also their appreciation for the support they received for this work from the Air Force Office of Scientific Research under Grant AFOSR 129-67, the National Aeronautics and Space Administration under Grant NsG-702/05-003-050 and the National Science Foundation under Grant NSF GK-2156.

Udine, September 1971   A. K. Oppenheim

CONTENTS

Contents

CHAPTER 1

EXPERIMENTAL APPARATUS AND RESULTS

1.1 Recording System

Laser oscillators, barely discovered a decade ago [1, 2] (*), can now be modulated by the use of Kerr or Pöckels cells [3, 4] to yield stroboscopic light pulses of only several nanoseconds in half-width at a frequency of an order of a mega-cycle per second [5] . Such a light source can be used in con-junction with a rotating-mirror camera combined with schlieren optics [6, 7] , a method that utilizes the fact that light is refracted by density gradients [8, 9] , to obtain cinematograph-ic records of explosion phenomena. The salient features of the apparatus are discribed in Fig. 1.1. The plane polarized light rays from the laser are expanded by a condenser lens and con-verted into an approximately 50 cm diameter beam of parallel light by means of the first schlieren lens or mirror. When the light beam passes through the test section, it is refracted by the density gradients associated with the gasdynamic flow field.

(*) Numbers in square brackets denote references listed at the end of the Chapter

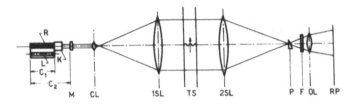

Fig. 1.1. Optical layout for stroboscopic schlieren system, C_1-light pump cavity, C_2-laser cavity, L-light pump, R-ruby rod, K-Kerr cell for Q-spoiling modulation of laser output, M-laser mirror, CL-condenser lens, 1 SL-front schlieren lens (or mirror), TS- test section, 2 SL- back schlieren lens (or mirror), P-schlieren aperture (quartz prism), F- polarizing filter, OL-objective lens, RP-receptor plane.

The light beam is then collimated by the second schlieren lens and passed through a quartz prism followed by a polarizing filter – a system which acts as an aperture that gradually cuts off the refracted light, forming, in effect, a so-called neutral wedge.

As explained in Chapter 2, the use of a neutral wedge in contrast to the conventional sharp edge is necessary in order to avoid the detrimental effects of diffraction, an inherent property of the laser light that mars the photographic resolution of the schlieren record. The remainder of the beam then enters a 120 cm diameter rotating mirror camera by means of which one obtains from each light pulse a frame of the cinematographic record of the changes in refractive index occurring in the test section. The rotating mirror, driven by a gas turbine, can turn at up 5,000 revolutions per second, yielding a writing speed on the film plane of approximately 2 cm per microsecond. The laser, its accessories, and the first schlieren mirror are shown in Fig. 1.2 (*), while the second schlieren mirror, the prism–polarized system and the rotating mirror camera are depicted in Fig. 1.3 (*).

(*) See plates pp. 35 ff.

1.2 Process Equipment

To study the propagation of explosion waves, an assortment of shock tubes and explosion vessels can be used, where, for optical observation, the test section must be equipped with ports fitted with optical glass windows.

Since the shock tube technique is so well established today that it can be considered as standard, while its fundamental features, especially with respect to applications in the study of explosion phenomena have been treated in the text of Oppenheim [10] , the description of these devices is omitted. Instead, presented here are the novel means associated with the use of explosion vessels which are especially suitable for the study of the properties of blast waves that are formed by a virtually instantaneous deposition of a finite amount of energy at a geometric point in the test gas. Among these are a steel vessel for the study of spherical explosions, Fig. 1.4 (*)thin vessels for the study of cylindrical explosions, Fig. 1.5 (*) and a lucite vessel for the study of spherical explosions under the influence of electric fields, Fig. 1.6 (*).

The traditional means used for the deposition of energy, i.e. electric sparks or exploding wires, are quite inadequate as a consequence of either the relatively long time they require to discharge the energy, which at the same time

_____

(*) See plates pp. 35 ff.

had to be deposited within a significant volume of the gas, or of the resulting metallic and electronic debris of the discharge that tend to contaminate the flow field. Thus, the ideal method for energy deposition is provided by focused laser irradiation. In the experiments to be described here this was achieved by the use of an American Optics neodymium laser, Fig. 1.7 (*), Q-spoiled by a Kodak 9740 saturable organic dye solution [11 – 13], whose concentration is adjusted to produce a single giant pulse of about 20 nanoseconds in half–width. This light pulse was then focused by means of a quartz plano–convex lens to cause an electric breakbown inside the explosion vessel or to sublimate a small amount of metal from a thin wire placed in the vicinity of the focus.

1.3 Experimental Results

Since the initial announcement by Terhune [14] that electrical breakdown can be induced in a gas by the focusing of a Q-spoiled laser pulse, the subject has received a great deal of attention [15 – 22] capped by the review paper of Meyerand [23]. Even though the exact breakdown mechanism is still a controversial matter, it is established that the initial presence of a few electrons is an important part of the process. At low densities, however, these free electrons are in short supply and it becomes then necessary to introduce a metallic

(*) See plates pp. 35 ff.

wire close to the focal point to provide electrons [24 - 26]
that are stripped off it by the electric field at the focus.
An example of such breakdown in air is shown in Fig. 1.8 (*) and
1.9. Fig. 1.8 is an open shutter photograph of the spark, ex-
hibiting its characteristic elongated shape [23, 27] , while
Fig. 1.9 is a set of cinematographic schlieren photographs,
taken at one microsecond intervals, recording the development
of the blast wave in the presence of a plasma ball that retains
an essentially constant radius after it becomes detached from
the front shock wave. The sequence in which the consecutive
frames are presented is, in all figures such as Fig. 1.9 (*),
from top to bottom, row by row from left to right. As it becomes evident from this record, although the 20 to 50 nanosecond duration of energy deposition can be considered to have
been virtually instantaneous in comparison to the 1 microsecond interval between the frames, the relatively long life of the plasma
ball tends to contaminate the medium as well as complicate the pattern
of energy release causing the resulting gasdynamic flow field for all
practical reasons to be analytically untractable.

       An alternative method that circumvents this dif-
ficulty is obtained by the use of a this wire placed in the path
of the laser beam at a small distance from the focal point.
There is no electric breakdown taking place in this case; instead
the irradiation of the wire results in a local sublimation as-

(*) See plates pp. 35 ff.

sociated with a small plume of vaporized metal [28 – 30] as shown in Fig. 1.10 (*). However, the accompanying schlieren record, Fig. 1.11 (*), taken in equimolar acetylene–oxygen mixture at 120 torr, indicates a virtual absence of such a plume. Apparently, in view of the small size of the laser beam close to the focal point, the wire acts as an extremely efficient heat sink for the vaporized metal caus- ing its rapid recondensation, and thus leaving the flow field rela- tively free of contaminants. One should note that in order to attain such conditions the amount of energy delivered by the laser must be adjusted so as not to produce a breakdown at the focal point in addi- tion to the hot spot on the wire. This situation is illustrated in Figs. 1.12 (*) and 1.13 (*). The analysis of the resulting phenomenon, name- ly the transition to detonation by the interaction of two spherical flames, even though it presents an interesting problem on its own right, is so complex that it pratically annihilates any advantages it may have in advancing the theory of explosions.

                To illustrate the scientific advantages offered by laser cinematography, the rest of this chapter will be de- voted to the demonstration how qualitative information on the salient phenomenological features of explosion processes can be derived on the basis of the experimental records obtained by this means. This will be done here without involving any quantitative analysis, especially since its fundamental aspects are available in the companion text [10] . The examples that

(*) See plates pp. 35 ff.

follow deal primarily with the development of detonation waves
and the generation of shock waves by flames - an essential step
in the process of transition to detonation. For the description
of other properties of explosions and detonation waves the reader is referred to the companion text [10] .

1.4 Transition to Detonation

The transition to detonation in long shock tubes
has been revealed primarily on the basis of experimental observations made by the use of laser-schlieren cinematography [31 -
34] . Upon ignition, a spheroidal laminar flame kernel that
propagates down the tube is generated. In due time, the surface
of the kernel becomes wrinkled and it starts accelerating and
emitting pressure waves which coalesce to form shock waves. The
flame then transits into a " tulip-shaped" turbulent structure
while vigorously emitting shock waves. The stage is thus set for an
"explosion in the explosion" resulting in the onset of a detonation
wave. The last stages of the development of this process are shown in
Fig. 1.14 (*) where it is apparent that two such secondary explosions
take place at 715 and 750 microseconds on the time scale.

No such clear-cut phenomena occur in the course of
transition to spherical detonation resulting from the instantaneous
deposition of a finite amount of energy, E_0, at a point - a process

(*) See plates pp. 35 ff.

which, as pointed out before, can be best achieved by an initiation associated with the use of a focused Q-switched laser pulse.

In general, two possibilities exist. At a low power density level, i.e. at low initial pressure, p_a , combined with a relatively small amount of initiation energy, E_0 , a blast wave is generated followed closely by a deflagration zone. The fast decay of the blast wave, however, causes the deflagration zone to decouple without inducing a transition to detonation. An example of this case has been shown here in Fig. 1.11, while its measured front trajectory diagram is given in Fig. 1.15. On the other extreme, at high power densities, i.e. for high values of p_a and E_0 , a Chapman-Jouguet detonation can be established almost immediately upon the initiation, as shown in Figs. 1.16 (*) and 1.17 (*). Between these extremes, a localized secondary explosion may occur ahead of the decoupled deflagration front [35, 36] generating thus

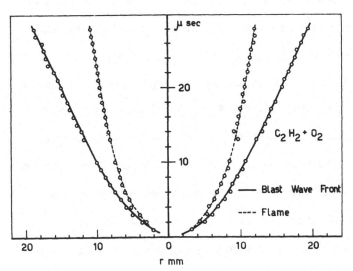

Fig. 1.15. Trajectories of shock and flame fronts of the wave system recorded in Fig. 1.11.

(*) See plates pp. 35 ff. and p. 19

a detonation wave as depicted in Figs. 1.18 (*) and 1.19 (*), or as in Figs. 1.20 (*) and 1.21 (*), a local separation may occur at a point in the established detonation wave that causes it to deteriorate into a decoupled system similar to that of Fig. 1.11.

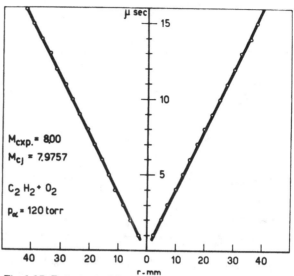

Fig. 1.17. Trajectory of detonation front recorded in Fig. 1.16. Distance measurements taken along horizontal lines passing through the center of explosion. The detonation reached its GJ velocity within the first microsecond.

1.5 Generation of Shock Waves by Spherical Flames

Under certain circumstances, it has been observed with the use of equimolar acetylene-oxygen mixtures that a decoupled flame kernel can suddenly start emitting a series of spherically symmetric shock waves. In addition to its

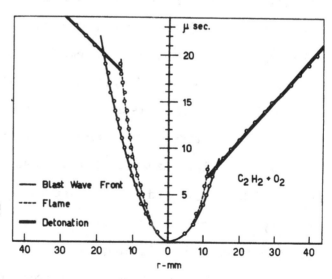

Fig. 1.19. Trajectories of wave fronts recorded in Fig. 1.18. Distance measurements taken along horizontal lines passing through the center of explosion.

(*) See plates pp. 35 ff. and p. 20

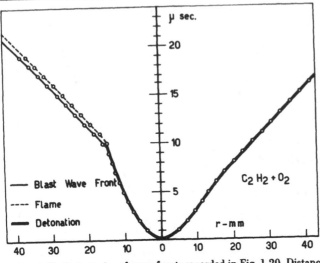

Fig. 1.21. Trajectories of wave fronts recorded in Fig. 1.20. Distance measurements taken along horizontal lines passing through the center of explosion.

importance to the study of combustion instability, this phenomenon reveals the possibility of the occurrence of an essential step in the transition to spherical detonation of the same kind, in fact, as that described in the previous section for the plane case.

Evidence of the ability of spherical flames to generate shock waves was first obtained for these mixtures in the range of initial pressures between 110 and 120 torr and room temperature. One such record is shown in Fig. 1.22 (*), another in Fig. 1.23 (*), while Fig. 1.24 (*) depicts the same phenomenon occurring while flame travels under the influence of an electric field.

Since, to our knowledge, no experimental evidence of such flame capabilities has ever been published before, it should be considered of interest to inquire whether it is due entirely to the combustion process rather than being associated with the effects of ignition.

The experimental records provide accurate data on the trajectories of the observed wave fronts in the time-

_____

(*) See plates pp. 35 ff.

space domain. The frequency of the Kerr cell acting as the Q-spoiler for the laser cavity is controlled directly by an oscillating crystal of a time-mark generator so that the repeatability of time intervals are accurate within at least a number of nanoseconds. At the intervals of 5 microseconds this corresponds to an accuracy of 1‰ . The position of the fronts in each frame were measured directly from the film by means of a spectroscopic microreader with a reading capability in microns. With our optical systems the ratio of the image on the film to object size is 0.13, for radii in millimeters this gives an accuracy better than 1%, yielding therefore the same order of magnitude for the precision in the velocity measurement.

The time-space diagram of the observed wave phenomena, deduced from the records of Fig. 1.22 and 1.23 are shown in Fig. 1.25 and 1.26 (*) respectively. The front velocities, evaluated by graphical differentiation of the trajectories of Fig. 1.25 are

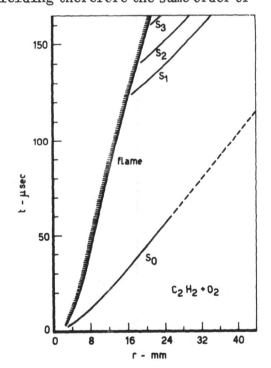

Fig. 1.25. Time-space trajectories of wave fronts recorded in Fig. 1.22.

(*) See plates pp. 35 ff.

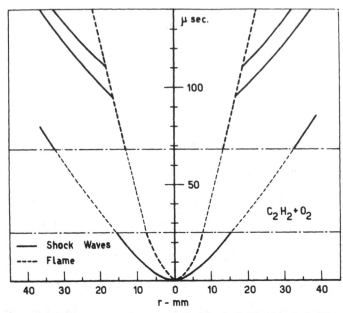

Fig. 1.26. Time-space trajectories of wave fronts recorded in Fig. 1.23. The thin dashed lines represent missing frames in the experimental records.

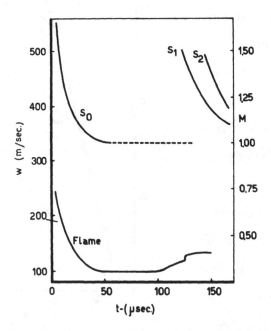

Fig. 1.27. Front velocities of wave fronts recorded in Fig. 1.22.

given in Fig. 1.27 (*). For general orientation, included in the latter is also the Mach number of the initial shock, M_s .

With the front Mach number of an order of 1.5, the initial blast wave must be treated as one corresponding to the case of a so-called " non zero counter-pressure" [37] . Properties of such decaying, constant energy blast waves have been evaluated recently by Korobeinikov and Chushkin [28] and the results, for three basic geometrical conditions (spherical, cylindrical and plane, corresponding respectively, to $j = 2, 1$ and 0) and a number of specific heat ratios, γ , were published in tabular form [39] . These results have been here graphically correlated, as shown on Figs. 1.28 (*) and 1.29 (*), demonstrating that all the front trajectories of the constant energy blast waves can be expressed, especially for small Mach numbers, by a single curve on the logarithmic time-space diagram, the differences associated with various values of γ and j being reflected only by appropriate shifts in the scales of coordinates.

The trajectory of shock S_0 has been matched with the front of a constant energy blast wave as depicted in Fig. 1.30 (*). The graph of Fig. 1.30 demonstrates also that trajectories of the shocks S_1 and S_2 , appearing later in front of the flame, have a completely different character. This is brought out clearer in the plot of the logarithmic front velocity $\mu = d\ln r/d\ln t$ given by Fig. 1.31(*). For a decaying,

(*) See plates pp. 35 ff.

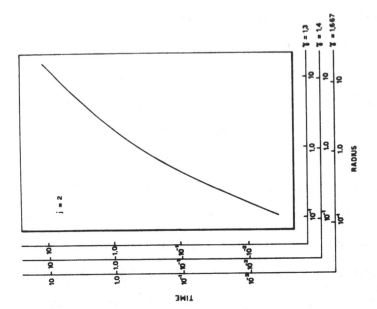

Fig. 1.29. Front trajectories in logarithmic time-space plane of constant energy blast waves for a point symmetrical flow field and $\gamma = 1.3$, 1.4, and 1.667 (based on data of Ref. 39).

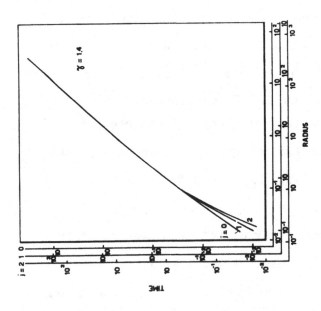

Fig. 1.28. Front trajectories in logarithmic time-space plane of constant energy blast waves for plane, line and point symmetrical flow field and $\gamma = 1.4$ (based on data of Ref. 39).

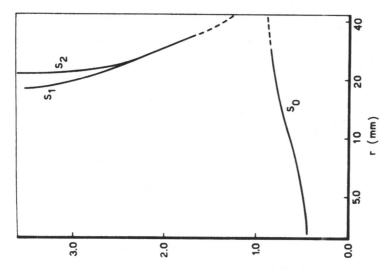

Fig. 1.31. Logarithmic velocities of experimentally observed shock fronts.

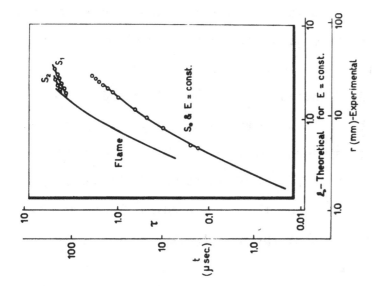

Fig. 1.30. Experimentally observed front trajectories in logarithmic time-space plane in comparison with the tra trajectory of a constant energy blast wave.

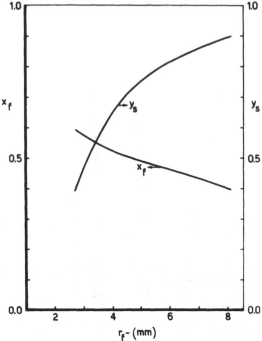

Fig. 1.32. Specification of the blast wave ahead of the flame in terms of the velocity parameter of the front $y \equiv 1/M_s^2$, and the flame position coordinate $x_f \equiv r_f/r_s$.

constant energy blast wave, this parameter varies from $\mu_s = 2/(j + 3)$ (or 2/5 for spherical geometry) to $\mu = 1$. On the other hand, shocks S_1 and S_2 start with μ's exceeding 3 and maintain values larger than unity throughout their existence.

From the above, one can conclude that shock S_0 is indeed a front of a constant energy decaying blast wave whose its strength is not affected by the heat released by the flame. The question still remains as to what extent is the motion of the flame front and of shocks S_1 and S_2 influenced by the flow field of the blast wave through which they evidently propagate. More specifically the portion of the blast wave under question is identified on Fig. 1.32 in terms of the shoch front parameter $y_s = 1/M_s^2$ and the corresponding flame coordinate $x_f = r_f/r_s$, subscript f referring to the flame and s to the shock front.

Some representative particle trajectories deduced from the tables of Korobeinikov and Chushkin in this por-

tion of the
blast wave are
shown on Fig.
1.33, and the
values of all
the gasdynam-
ic parameters
of the blast
wave immediate_

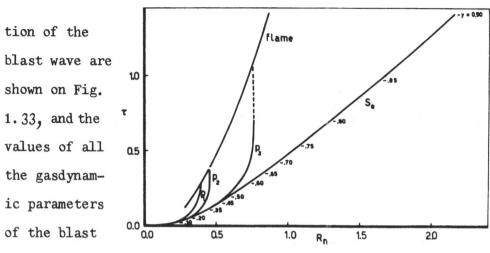

Fig. 1.33. Representative particle trajectories of the blast wave in the time-space domain.

ly ahead of the flame front, without any account taken of the
flame generated flow field, are given in Fig. 1.34 (*).

As it appears from the latter, after the initial
transient stage, the flow field of the blast wave ahead of the
flame attains, considerably before the appearance of shock S_1 ,
a pratically steady state of zero particle velocity, undisturbed
pressure, and the density approaching slowly its initial value.
Thus, for all intents and purposes, the transient processes of
the initial blast wave associated with ignition can be neglected, except for the small increase in the temperature reflected
by the density curve. In particular, one may conclude that the
creation of shocks S_1 and S_2 is therefore, due entirely to the effects of the flame generated flow field.

Thus, the capability of a sherical flame to generate shock waves has been experimentally demonstrated. The

(*) See p. 28

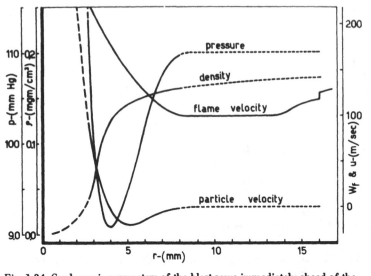

Fig. 1.34. Gasdynamic parameters of the blast wave immediately ahead of the flame, without any account taken of the flame generated flow field.

unique records of the flame and of the shock fronts associated with it have been subjected to a through analysis in order to establish the fact that the influence of the blast wave formed by the ignition process on the observed phenomena is negligible.

It should be borne in mind, that the analysis presented here does not take into account the flow generated by the flame. Thus the gasdynamic parameters of state evaluated immediately ahead of the flame, as shown in Fig. 1.34 represent only their components due to the blast wave. Their actual values should be different as a consequence of the flame generated flow field. The analysis of such effects is outside the scope of this monograph. One has to be satisfied thus with the observation that, although the flame velocity in Fig. 1.34 is, in effect, the sum of the burning speed and the particle velocity due to the action of the flame, it can be shown that, under the operating conditions of the experiment, the latter takes up most of the value observed experimentally, the particle velocity be-

ing of an order of 100 m/sec and the burning speed just about
3 m/sec . The increase in the flame velocity recorded in Fig.
1. 34 just before the appearance of shock S_1 can be ascribed,
therefore, most probably to the fact that the flame became turbulent, the transition being due, in turn, to the fact that
the flow field ahead of the flame, being associated with a
relatively high velocity, was itself turbulent.

References

[1] Schlow, A.L. and Tawnes, C.H. " Infrared and Optical Lasers", _Phys. Review_, 112, pp. 1940-1949, 1958.

[2] Maiman, T.H. " Stimulated Optical Radiation in Ruby", _Nature_, 187, p. 493, 1960.

[3] Maiman, T.H. " State of the Art Devices", _Optical Masers_, (Fox, J. ed.), Polytechnic Press of Polytechnic Institute of Brooklyn, N.Y., 1963.

[4] Wagner, W.G., Lengyel B.A. " Evolution of the Giant Pulse in Laser", _J. Appl. Phys._, 34, p. 2040, 1963.

[5] Lo, C.C., " A2 - MH_z 8 - KV Pulser for High-Speed Strobo-scopic Photography", _Lawrence Radiation Labora-tory_, UCRL - 19248, University of California, Berkeley, 1969.

[6] Oppenheim, A.K., Urtiew, P.A. and Weinberg, " On the Use of Laser Light Sources in Schlieren-Interfero-meter Systems", _Proc. Roy. Soc._, A291, 279-290, 1966.

[7] Hecht, G.J., Steel, G.B. and Oppenheim, A.K., " High-Speed Stroboscobic Photography Using a Kerr Cell Modulated Laser Light Source", _ISA Trans._ Vol.5, N°2, pp. 133-138, April 1966.

[8] Weinberg, F.J., _Optics of Flames_, 251 pp., Butterworths, London, 1963.

[9] Soloukhin, R.I., _Udarnye Volny i Detonatsia v Gazakh_ (Shock Waves and Detonation in Gases), 175 pp., Gos. Izd. Fiz. Mat. Literatury, Moscow, 1963; (Transl. by B.W. Kuvshinoff, Mono Book Corp., Baltimore, 1966); AMR, Vol. 18, 1965, Rev. 342.

[10] Oppenheim, A.K., Introduction to Gasdynamics of Explosions, International Centre for Mechanical Sciences (CISM), Udine, Italy, 1971.

[11] Smith, W.V. and Sorokin, P.P. The Laser, 498 pp., McGRaw-Hill Book Co., N.Y., 1966.

[12] Cross, L.A. and Cheng, C.K., "Generation of Giant Pulses from a Neodymium Laser with an Organic-Dye Saturable Filter", J. Appl. Phys., Vol. 38, N°5, pp. 2290-2294, April 1967.

[13] Skeen, C.H. and York, C.M., "The operation of a Neodymium Glass Laser Using a Saturable Liquid Q-Switch", App. Opt., Vol. 5, N°9, pp. 1463-1464, 1966.

[14] Terhune, R.W., "Non-Linear Optics", Bull. Am. Phys. Soc., Vol. 8, p. 359, 1963.

[15] Tomlinson, R.G., "Gas Breakdown Criterion for Pulsed Optical Radiation", Proc. of the IEEE, Vol. 52, N°6, pp. 721-722, 1964.

[16] Bebb, H.B. and Gold, A., "Multiphoton Ionization of Rare Gas and Hydrogen Atoms", Phys. of Quantum Electronics, McGraw-Hill Book Co., N.Y., pp. 489-498, 1965.

[17] Haught, A.F., Meyerand, R.G. Jr. and Smith, D.C., "Electrical Breakdown of Gases by Optical Frequency Radiation", Phys. of Quant. Elec., McGraw-Hill Book Co., pp. 509-519, 1965.

[18] Tomlinson, R.G., Damon, E.K. and Buscher, H.T., "Breakdown of Nobel and Atmospheric Gases by Ruby and Neodymium Laser Pulses", Phys. Quant. Elect., McGraw-Hill Book, pp. 520-526, 1965.

[19] Minck, R.W. and Rado, W.G., "Investigation of Optical Frequency Breakdown Phenomena", Phys. Quant.

Elec., McGraw-Hill Book Co., pp. 527-537, 1965.

[20] Phelps, A.V., " Theory of Growth of Ionization During
Laser Breakdown", Phys. Quant. Elec., McGraw-
Hill Book Co., pp. 538-547, 1965.

[21] Peressini, E.R., " Field Emission from Atoms in Intense
Optical Fields", Phys. Quant. Elec., McGraw-
Hill Book Co., pp. 499-508, 1965.

[22] Tomlinson, R.G., " Atmospheric Breakdown Limitations to
Optical Maser Propagation", Radio Science J. of
Research, Vol. 69, pp. 1431-1433, 1965.

[23] Meyerand, R.G. Jr, " Laser Plasma Production - A New
Area of Plasmadynamics Research", AIAA J., Vol.
5, N°10, pp. 1730-1737.

[24] Lee, John H. and Knystautas, R., " Laser Spark Ignition
of Chemically Reactive Gases", AIAA J., Vol. 7,
pp. 312-317, 1969.

[25] Knystautas, R., " An Experimental Study of Spherical
Gaseous Detonation Waves", McGill University
MERL, Report 69-2, Montreal, Canada, 1969.

[26] Bach, J.J., Knystautas, R., and Lee, J.H., " Direct Initiation of Spherical Detonation in Gaseous Ex-
plosives", Twelfth Symposium (International) on
Combustion. The Combustion Institute, Pittsburgh,
Pa., pp. 853-864, 1969.

[27] Weinberg, F.J. and Wilson, J.R., " A Preliminary Investi-
gation of the Use of Focoused Laser Beams for
Minimum Ignition Energy Studies", Proc. Roy.
Soc. Lond., A321, pp. 41-52, 1971.

[28] Ready, J.F., " Development of Plume of Material Vaporized
by Giant Laser", Appl. Phys. Letters, Vol. 3,
N°1, p. 11, 1963.

[29] Honig, R.E., " Laser-Induced Emission of Electrons and

Positive Ions from Metals and Semi–Conductors",
Appl. Phys. Letters, Vol. 3, N°1, pp. 8–11, 1963.

[30] Honig, R.E. and Woolston, J.R., " Laser–Induced Emission
of Electrons, Ions, and Neutral Atoms from Solid
Surfaces", Appl. Phys. Letters, Vol. 2, N°7,
pp. 138–139, 1962.

[31] Oppenheim, A.K., Manson, N., and Wagner, H.Gg., " Recent
Progress in Detonation Research", AIAA Journal,
Vol. 1, N°10, pp. 2243–2252, 1963.

[32] Oppenheim, A.K., " Novel Insight into the Structure and
Development of Detonation", Astronautica Acta,
Vol. 11, pp. 391–400, 1965.

[33] Urtiew, P.A., and Oppenheim, A.K., " Detonative Ignition
Induced by Shock Merging", Eleventh Symposium
(International) on Combustion, pp. 665–670, 1967.

[34] Lee, J.H., Soloukhin, R.I., and Oppenheim, A.K., " Current Views on Gaseous Detonation", Astronautica
Acta, Vol. 14, pp. 565–584, 1969.

[35] Lee, J.H., Lee, B.H.K., and Knystautas, R., " Direct
Initiation of Cylindrical Gaseous Detonations",
Physics of Fluids, Vol. 9, pp. 221–222, 1966.

[36] Lee, J.H., and Knystautas, R., " Laser Spark Ignition
of Chemically Reactive Gases", AIAA Journal,
Vol. 7, N°2, pp. 312–317, 1969.

[37] Sedov, L.I., Similarity and Dimensional Methods in Mech-
anics, (Moskow: Gastekhizdat) 4th ed., 1957,
(English translation, ed. M. Holt, New York:
Academic Press, XVI 363 pp., 1959).

[38] Korobeinikov, V.P., and Chushkin, P.I., " Plane Cylin-
drical and Spherical Blast Waves in a Gas with
Counter-Pressure", Proc. V. A. Steklov Inst. of
Math. (in " Non–Steady Motion of Compressible

Media Associated with Blast Waves" edited by
L.I. Sedov), pp. 4-33, Izdatel'stvo " Nauka" ,
Moscow, 1966).

[39] Korobeinikov, V.P., Chushkin, P.I., and Sharovatova, K.V.,
" Gasdynamic Functions of Point Explosions" ,
Computer Center of the USSR Academy of Sciences,
Moscow, 1969.

PLATES FOR CHAPTER 1

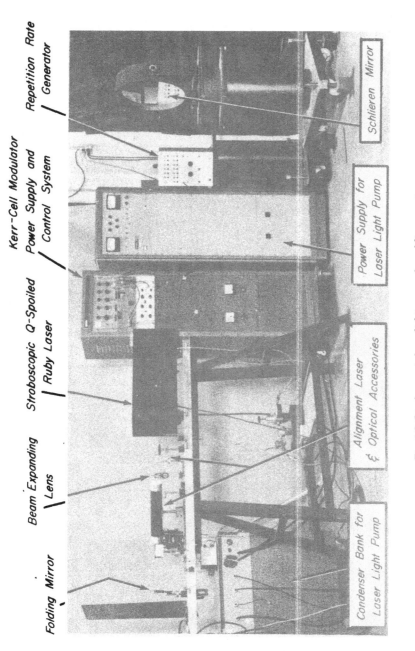

Fig. 1.2. Stroboscopic laser light source for schlieren system.

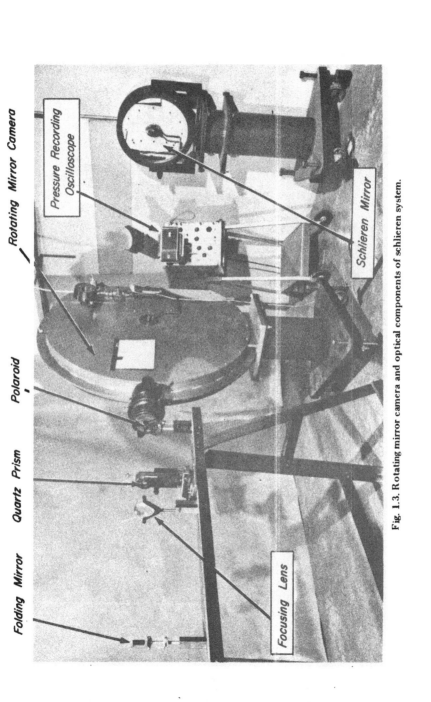

Rotating Mirror Camera

Pressure Recording
Oscilloscope

Schlieren Mirror

Polaroid

Quartz Prism

Folding Mirror

Focusing Lens

Fig. 1.3. Rotating mirror camera and optical components of schlieren system.

Fig. 1.4. Steel vessel for the study of spherical blast waves.

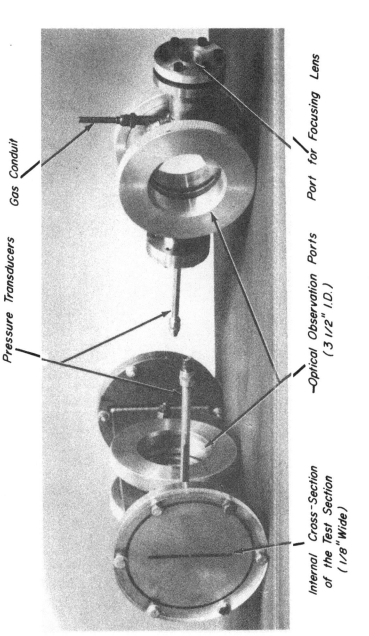

Pressure Transducers   Gas Conduit

Port for Focusing Lens

Optical Observation Ports
(3 1/2" I.D.)

Internal Cross-Section
of the Test Section
(1/8" Wide)

Fig. 1.5. Vessels for the study of cylindrical blast waves.

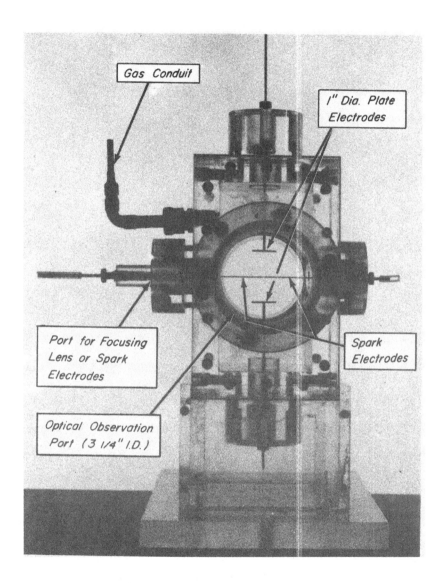

Fig. 1.6. Lucite vessel for the study of spherical blast waves under the influence of electric fields.

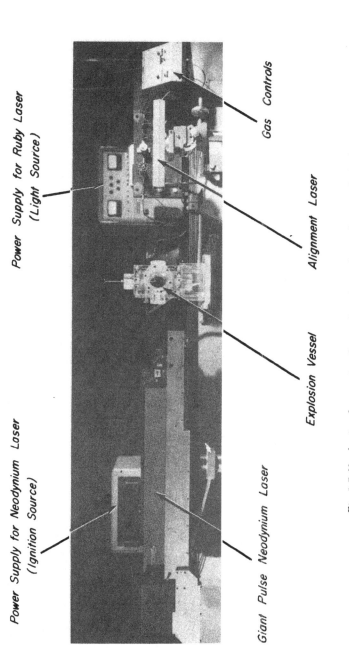

Power Supply for Ruby Laser
(Light Source)

Gas Controls

Alignment Laser

Power Supply for Neodynium Laser
(Ignition Source)

Explosion Vessel

Giant Pulse Neodynium Laser

Fig. 1.7. Neodymium laser system for use in a giant pulse mode as a point ignitor.

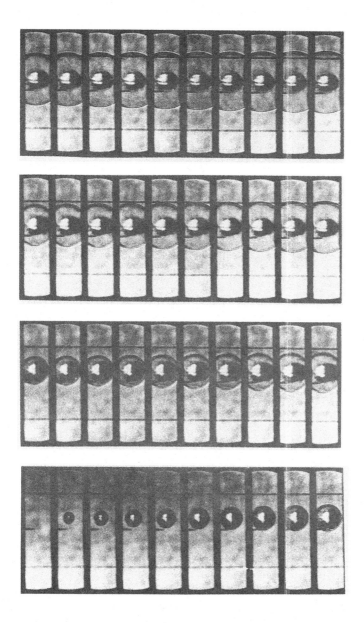

Fig. 1.9. Stroboscopic laser-schlieren photographs of a blast wave generated in air by laser breakdown. The initial pressure was 1 atm. The time interval between frames, in the sequence from top to bottom and left to right, was 1 microsecond. The 2 vertical marking lines were set at a distance of 2 inches.

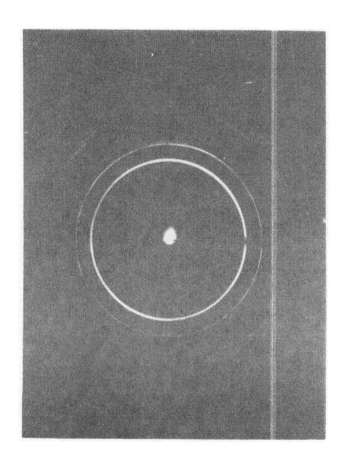

Fig. 1.10. Open shutter photograph of a neodymium laser giant pulse breakdown in air at 120 torr. Laser beam was focused on a metallic wire.

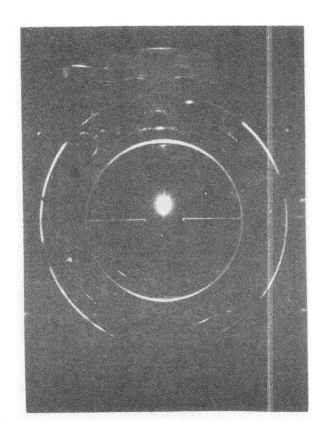

Fig. 1.8. Open shutter photograph of focused neodymium laser giant pulse breakdown in air at 1 atm.

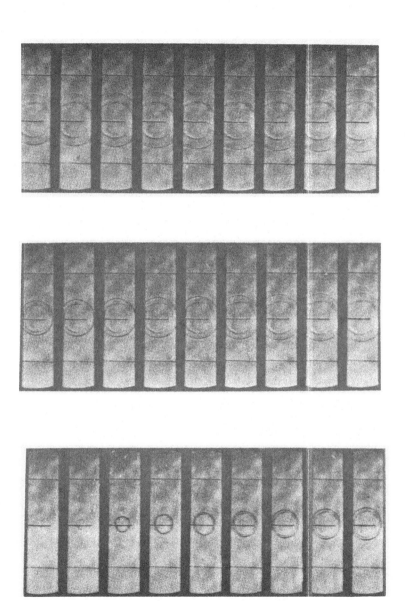

Fig. 1.11. A stroboscopic laser-schlieren photograph of a decoupled shock wave-flame system. The medium was equimolar oxygen-acetylene mixture at an initial pressure of 120 torr. Ignition performed by focusing a neodymium laser giant pulse on a metallic wire. The time interval between frames, in the sequence from to bottom and left to right, was 1 microsecond.

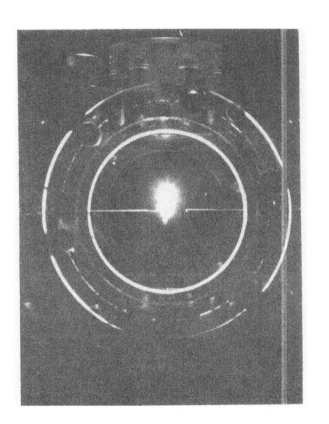

Fig. 1.12. Open shutter photograph of a double breakdown of a neodymium laser giant in the presence of a metallic wire.

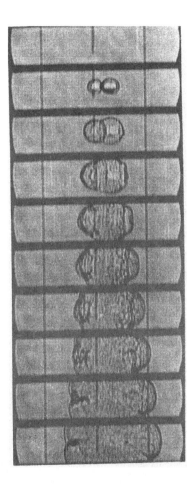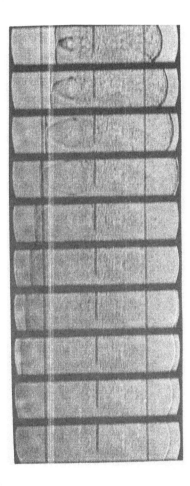

Fig. 1.13. A stroboscopic laser-schlieren photograph of transition to detonation resulting from the interaction between two spherical flames produced by a double breakdown of a neodymium laser giant pulse in the presence of a metallic wire. The medium was equimolar acetylene-oxygen mixture at an initial pressure of 120 torr. The time interval between frames, in the sequence from top to bottom and left to right, was 1 microsecond.

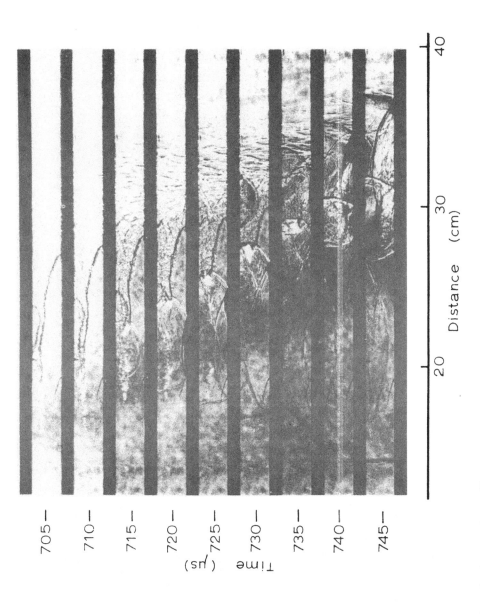

Fig. 1.14. Stroboscopic laser-schlieren photographs of the onset of detonation as a consequence of secondary explosions in the exploding gas. The medium was hydrogen-oxygen mixture ignited by a spark near the closed end of a 1" x 1-1/2" cross-section tube.

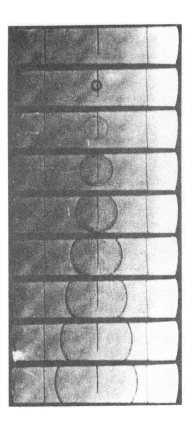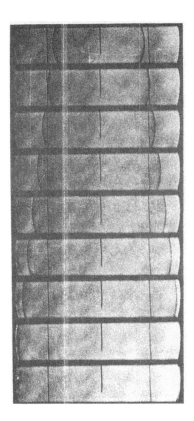

Fig. 1.16. Stroboscopic laser-schlieren photograph of the direct initiation of detonation due to the deposition of energy by focusing a neodymium laser giant pulse on a metallic wire. The medium was equimolar oxygen-acetylene mixture initially at a pressure of 120 torr. The time interval between frames, in the sequence from top to bottom and left to right, was 1 microsecond.

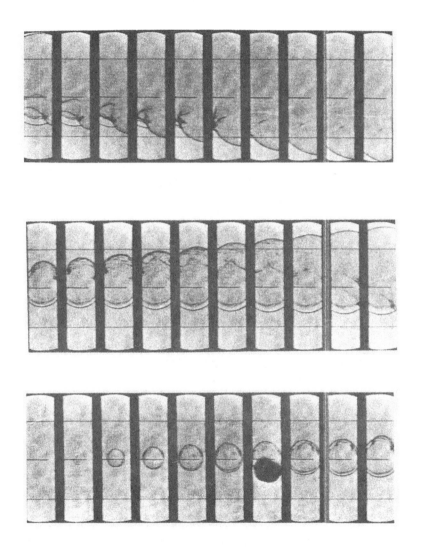

Fig. 1.18. Stroboscopic laser-schlieren photographs of the generation of detonation by a spherical flame. The time interval between frames, in the sequence from top to bottom and left to right, was 1 microsecond. The two vertical marking lines were set at a distance of 2 inches. The medium was an equimolar acetylene-oxygen mixture initially at a pressure of 100 torr and room temperature. Ignition was performed by the giant pulse of the neodymium laser.

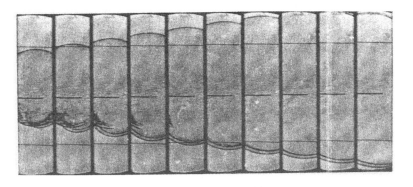

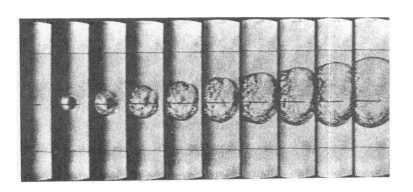

Fig. 1.20. Stroboscopic laser-schlieren photographs of the decoupling of a detonation wave. The time interval between frames, in the sequence from top to bottom and left to right, was 1 microsecond. The two vertical marking lines were set at a distance of 2 inches. The medium was an equimolar acetylene-oxygen mixture initially at a pressure of 100 torr and room temperature. Ignition was performed by the giant pulse of the neodymium laser.

a

Fig. 1.22. Stroboscopic laser-schlieren photographs of the generation of shock waves by a spherical flame. The time interval between frames, in the sequence from top to bottom, and from left to right, is 5 microseconds. The outside diameter of the field of view is 9 cm. The medium was an equimolar acetylene-oxygen mixture, initially at a pressure of 110 torr and room temperature. Ignition was performed by the giant pulse of the neodymium laser.

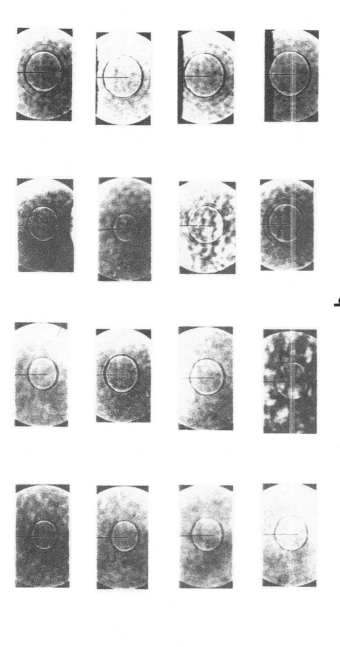

b

Fig. 1.22. Stroboscopic laser-schlieren photographs of the generation of shock waves by a spherical flame. The time interval between frames, in the sequence from top to bottom, and from left to right, is 5 microseconds. The outside diameter of the field of view is 9 cm. The medium was an equimolar acetylene-oxygen mixture, initially at a pressure of 110 torr and room temperature. Ignition was performed by the giant pulse of the neodymium laser.

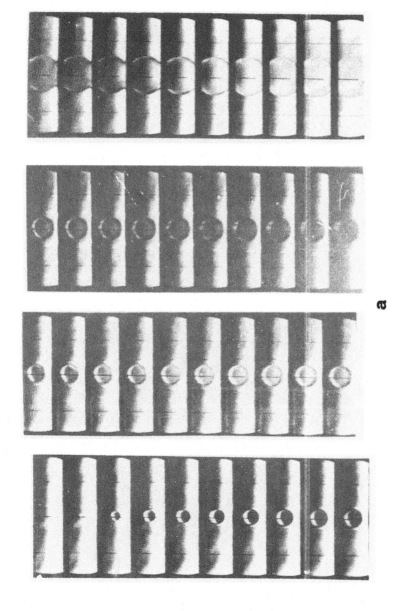

a

Fig. 1.23. Stroboscopic laser-schlieren photographs of the generation of shock waves by a spherical flame. The time interval between frames, in the sequence from top to bottom and from left to right, is 1 microsecond. The medium was an equimolar acetylene-oxygen mixture, initally at a pressure of 120 torr and room temperature. Ignition was performed by the giant pulse of the neodymium laser.

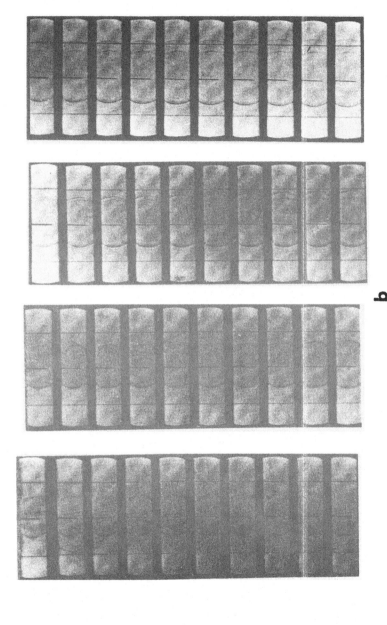

b

Fig. 1.23. Stroboscopic laser-schlieren photographs of the generation of shock waves by a spherical flame. The time interval between frames, in the sequence from top to bottom and from left to right, is 1 microsecond. The medium was an equimolar acetylene-oxygen mixture, initally at a pressure of 120 torr and room temperature. Ignition was performed by the giant pulse of the neodymium laser.

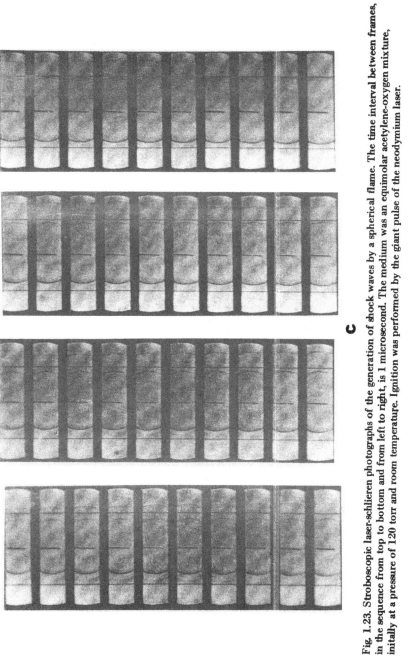

c

Fig. 1.23. Stroboscopic laser-schlieren photographs of the generation of shock waves by a spherical flame. The time interval between frames, in the sequence from top to bottom and from left to right, is 1 microsecond. The medium was an equimolar acetylene-oxygen mixture, initally at a pressure of 120 torr and room temperature. Ignition was performed by the giant pulse of the neodymium laser.

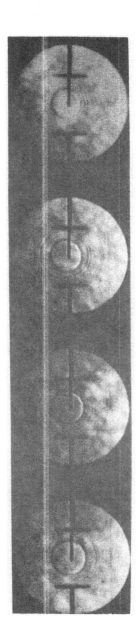

Fig. 1.24. Stroboscopic laser-schlieren photographs of the generation of shock waves by a spherical flame in the presence of an electric field taken in the lucite vessel in Fig. 1.6. The time interval between frames, in the sequence from top to bottom, first row followed by the second, is 5 microseconds. The medium was an equimolar acetylene-oxygen mixture, initially at a pressure of 100 torr and room temperature. The electric field applied was 4.0 KV.

CHAPTER 2

OPTICS

2.1 Introduction

As it is well known today, the advent of lasers revolutionalized the scientific optical recording technology. The most popular in this respect is holography – a technique made pratically possible as a consequence of the appreciable coherence length of the laser light [1] (*). It is of interest to note, however, that such exploitation of the advantages accrued by the use of lasers as light sources is by no means associated with the introduction of new principles. The fundamentals of holography, for instance, have been laid down by Gabor [2] long before laser was discovered. Thus the real nature of the revolution brought about by the use of lasers as light sources for optical observations is associated primarily with the improvements they introduced to experimental techniques whose basic features were known before.

The most important advantage associated with the use of lasers for the study of explosions is derived from

(*) Numbers in square brackets denote references listed at the end of the Chapter.

their exceedingly high brightness, particularly when used in short duration flashes which are required in order to obtain a high frequency cinematographic effect. Although for this purpose one must use a solid state (ruby)laser, the basic work required for the development of such techniques is carried out using a gas laser, on account of its convenience for setting-up procedures. With the exception of the rate of energy release, the properties of the two sources are closely similar, even to the wave length of the radiation emitted which is 6943 Å for the ruby laser and 6328 Å for the helium-neon gas laser.

The unique attributes of laser light which demand special consideration when it is used as a source for the observation and recording of refractive index fields associated with explosion phenomena are:

(1) its exceptional parallelity and coherency;

(2) its almost perfect monochromaticity;

(3) its inherent polarization.

The first of these leads to the formation of exceedingly small images at focal points. Their finite extent is largely that of their diffraction patterns caused by confining apertures. This enhances fringe formation and consequently allows more freedom in interferometry. However, the minute images tend to be damaging to optical components placed at the foci and lead to troublesome diffraction effects in the methods of geometrical optics.

The second makes possible interference between beams of very large path difference and thereby provides the additional facility of exceedingly high fringe order in interferometry. It also minimizes all dispersive effects, including chromatic aberration, and allows the use of prisms as optical elements in schlieren systems.

The third provides the facility for a variable attenuation of beams when a polarizing screen is used at varying angles to the incident direction of polarization. This is not accompanied by the troublesome diffraction effects associated with confinement by apertures which are usually employed for this purpose in schlieren systems.

Properties of laser light pose special problems, as well as presenting special opportunities. There are examined here in relation to schlieren recording, deflection mapping, shadowgraphy, and interferometry, leading to the development of a number of optical systems which are all simple and inexpensive adaptions of the conventional parallel-beam schlieren arrangement, and fulfill these various functions.

The fundamental background for this chapter has been provided primarily by the contributions of Weinberg described in Refs. [3] and [4] . The reader interested in the application of pulsed lasers to holographic interferometry is referred to the publications of Wuerker and his associates [5, 6] for specific features of laser interferometry to the numerous

papers of Tanner [7 - 10] , and, especially with respect to the
excellent spatial resolution attained by the use of scatter plates,
to the article of Gates [11] .

2.2 Schlieren System

In order to distinguish it clearly from other
methods based on ray deflections by refractive index gradients,
and in accordance with previous usage [12] , the term " shlieren
system" is here confined to that in which the receptor is optically conjugate with the test space and deflected regions are
marked by changes in illumination or color, but not by any distortion of the image. This will be considered first, because
the optical layouts, throughout this exposition are based on a
conventional parallel-beam schlieren system. The particular arrangement referred to here employs mirrors 46 cm in diameter and
4 meters focal length, and its laboratory lay-out is shown in

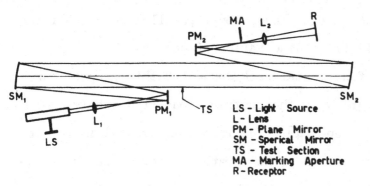

Fig. 2.1. Basic schlieren system .

Fig. 2.1.
The variation of
image illumina-
tion, I , with
angle of deflec-
tion, θ , in-
duced in the test
space, is expres-

sed in terms of the variation of light transmission, t , with

position, p , at the focus of the second schlieren mirror as

follows [12] :

$$\frac{\Delta l}{I_0} = \frac{\Delta t}{t_0} = \frac{1}{t_0}\left(\frac{dt}{dp}\right)\Delta p = \frac{1}{t_0}\left(\frac{dt}{dp}\right)F\theta\cos\phi \qquad (2.1)$$

where F is the focal length of the second schlieren mirror and

ϕ is the angle between Δp and the direction of deflection,

whilst subscript 0 refers to $\theta = 0$

The particular circumstance which arises when a

laser beam is used with the conventional knife-edge is that its

image at the focus of the schlieren mirror constitutes virtual-

ly a point of monochromatic light in the absence of deflections.

Thus, on geometric theory, dt/dp is infinite, and so is the

contrast-sensitivity of the system. The range, on the other

hand, is zero; i.e., the angle over which changes in deflection

are marked by changes in illumination is infinitesimal, and the

system should be " on-off" in operation. The above geometric

theory, however, is only an approximation which, in particular,

neglects all diffractions effects. The circumstances of infi-

nite sensitivity are precisely those under which the approximation

breaks down, as it has been, indeed, demonstrated in the various

studies of the performance of schlieren systems examined in

terms of wave theory that are available in the literature [12 –

17] . It is of interest to note that even the elementary theory

[14] of an infinitesimal source (except for the fact that it

refers to a line rather than a point) emphasizes, in this respect, the particular properties of laser light. It indicates
[15] that diffraction leads to the appearance of increased marking around all boundaries restricting the aperture, giving rise
to illumination also outside these boundaries and rounding the
contours of all deflection profiles. These features become more
and more prominent as sensitivity is increased (Ref. [12] , pp.
125-128) and the " on-off" response is never attained. Physical
ly this stems from the fact that at the second focus the knife-
edge interacts with the fringes of a diffraction pattern rather
than with a geometrical point of light.

      Figure 2.2 (*) is a sequence of schlieren photographs
taken by a flashing ruby laser which records the propagation of
a shock wave followed by a turbulent flame in a 2.54 × 3.81 cm
rectangular cross-section tube. The knife edge was parallel to
the tube axis. In contrast to the predictions of the geometric
theory for the infinitesimal source, a gradual (rather than "on-off")
variation in illumination (but not obeying Eq. 2.1!) has been obtained and the diffraction patterns surrounding the shock wave
extend well beyond the confining apertures of the test field.

      While there is no way of circumventing the wave
nature of light, the geometrical theory can be reinstated – at
least as an approximation – and diffraction effects can be min-
imized by replacing the knife-edge by a neutral wedge with a

(*) See plates pp. 111 ff.

sufficiently gradual dt/dp . This allows range and sensitivity to be suited to the particular deflection field under study
even with an hypothetical point of light, and removes one straight
edge which would otherwise strengthen the diffraction pattern.

The effect of this modification is illustrated in
Fig. 2.3 (*). It shows three records obtained under identical conditions except for the marking aperture. In the case of the
knife edge demonstrated by Fig. 2.3a the variation is almost
entirely in the form of fringes, which arise as successive diffraction maxima are cut off by the knife edge. Figure 2.3b,
taken with a photographically produced neutral wedge, whilst
retaining some of the inevitable diffraction effects, has the
normal appearance of a schlieren record with finite dt/dp , whose
gradual variations can be interpreted in accordance with Eq. (2.1).

For the gas laser photographically produced neu-
tral wedges are adequate. In the case of the ruby flash however,
energy absorbed at the focus, particularly in the absence of a
strong and extended deflection field to distribute light, dam-
age the film. Hence a thermally more robust arrangement is required. The use of metal films on a quarts base is an obvious
possibility; for practical reasons, however, an alternative
method due to Weinberg [4] , exploiting the polarization of the
laser light is more convenient.

The method is based upon the principle that a

(*) See plates pp. 111 ff.

prism of quartz, or other suitable material which rotates the
plane of polarization, produces a final direction of polarization depending on the thickness traversed and, hence, on the
position of the incident beam (*). Thus plane-polarized rays of
approximately $6500\,\overset{\circ}{A}$ in wavelengths entering a 30° quartz prism
1 cm apart will emerge with their planes of polarization prac-
tically perpendicular to one another. When such a prism is placed
at the focus of the second schlieren mirror, and is followed by
a sheet of polaroid anywhere beyond it, variation from transmission to extinction will result for an angle of deflection, θ^*,
in the test zone   given by

$$(2.2) \qquad\qquad\qquad \theta^* = \frac{p^*}{f}$$

where f is the focal length of the second schlieren mirror and
   p^* is the distance along the prism within which the angle of
rotation changes by 90°. In this case dt/dp in Eq. (2.1) follows
cosine law. Figure 2.3c is a schlieren record of the test object
of Figs. 2.3a and 2.3b, taken by this method. The quartz prism
was of 30° and 90° with the optical axis parallel to its short-
est face. The incident beam was made parallel to this direction
and its undeflected position situated half way between regions
of transmission and of complete extinction.

   All the photographic records presented in Chapter

(*) For principle of polarization by crystals see Appendix A.

1 have been obtained by means of the quartz prism - polaroid
method. In particular Fig. 1.14, representing a photograph of
essentially the same phenomena as Fig. 2.2, demonstrates striking-
ly the improvements one obtains by its use.

This device has a number of interesting properties. With the undeflected beam at a maximum or minimum in
transmission (arranged either by moving the prism or rotating
the polaroid filter), symmetrical marking results in the two
directions in the plane of the prism. The device then behaves
as a confining slit rather than a knife edge. When prisms of
larger angles are used, the illumination is made to undergo
successive extinctions and maxima over the range of deflections
produced in the test space, resulting in "Ronchi" fringes. It
will be apparent that when introduced elsewhere in the system,
this device can be used for deflection mapping by producing
fringes similar to those discussed in the next section.

Finally one should note that the introduction
of a prism results in a deflection of the whole beam, producing a
distortion of the image. For the records of Chapter 1, this has
been actually compensated for by the use of a second complementary neutral prism. However one may also correct this drawback
by placing the receptor at a proper angle to the original direction of the optical axis.

2.3 Deflection Mapping

Since it is more accurate to measure a shape than an intensity of blackening on a photographic emulsion, extensive use has been made (Ref. [12] , pp. 173ff) of recording the distortion of shapes (usually straight lines) by the refractive index field under study. The most obvious method of producing such reference marks on the wave-front is by using a grid of alternate transparent and opaque strips [18, 19, 20] .

The accuracy is limited by the relative magnitude of the angle of deflection to be measured and the angle of uncertainty in reading the record, caused by the spread of the diffraction patterns of each slit. This interacts with the determinacy of the position of origin of the deflection; the narrower a slit is made in order to define this position accurately, the wider its diffraction pattern and the greater the inaccuracy in reading the magnitude of deflection.

These difficulties are unimportant for large deflections extending over appreciable distances, but become serious for small angles ($\approx 10^{-3}$ rads). Fortunately, the requirements for accuracy in interferometry are exactly complementary and there is adequate overlap between the ranges of the two methods (Ref. [12] , p. 236).

The limited resolving power is most obvious in

the case of simple shadow recording of a distorted grid. As

sensitivity is increased (e.g., by increasing separation of

the receptor) the displacement increases in proportion to the

width of the diffraction pattern. Some alleviation can be pro-

duced by various lens systems [21, 22, 23]). In these, the grid

and test space are at different positions and hence they are

portrayed at different magnifications. The simple proportion-

ality between displacement and width of diffraction pattern on

the record is therefore broken, and an additional degree of

freedom becomes available.

Superficially, the optimum system might appear

to be that in which the marking screen is conjugate with the

receptor, while the test space is not, and, in the absence of

deflections, this indeed results in the disappearance of dif-

fraction patterns on the record. However, when large deflec-

tions occur, the image is broken into fringes caused by the

superposition of diffracted wave facets which differ in phase

because of different optical paths following deflection. The

most legible records, on the contrary, are often produced with

the test space conjugate with the receptor [21] . The theo-

ry converting distortions into angles of deflection is dexcribed

in reference [12] on pp. 182ff.

A fundamental method of improving the resolv-

ing power is to use a " physical optics" principle for mark-

ing the wave-front rather than the " geometrical" one of casting

shadows. The use of half-wave steps [24, 25] results in consider
able improvement. In this, a grid of relatively sharp lines is
produced by interposing into the beam a glass plate coated with
strips of transparent material, half a wavelength thick. Along
every plane which contains the half wave step at each edge destructive interference between the parts of the wave front that
are out of phase results in a dark line.

An alternative method, to which the laser light
source lends itself readly, is the use of evenly spaced straight
interference fringes as "reference marks on the wave front."
The records obtained in this manner have every appearance of
interferograms, but are in fact produced by deflection-induced
distortion of fringes.

The optical path difference, giving rise to interference, takes place not in the test space, but in an element
specially introduced for this purpose. Almost any glass slab will
serve in a laser system. In the parallel part of the beam, a
partially silvered wedge, of a Fabry-Perot etalon, may be used
to produce "fringes of constant thickness." Such an expense,
however, seems quite unnecessary, as any approximately plane
piece of glass (e.g. a microscope slide) will produce sharp
fringes, if introduced at a sufficient inclination near one of
the foci. This is due simply to multiple reflections at such an
oblique angle that reflectivity becomes appreciable.

Figures 2.4a and 2.4b show such fringes produced

using a glass slab near the first focus. The purpose of Figs.
2.4 is to compare the invariability of such fringe marking with
the diffraction spread of a Ronchi grid. Figure 2.4a shows a
fringe pattern of spacing equal to that of the grid (5 lines/cm)
of Fig. 2.4c which was recorded at the position in the parallel
beam at which the grid was subsequently introduced. Figures 2.4b
and d contrast the fringe and grid records taken in the parallel beam at a distance of 2.5 meters behind the first position.
Whereas the marking produced by the plate remain almost invariant, those of the Ronchi grid became completely illegible.

The use of the principle for deflection mapping
is illustrated by Figs. 2.5a and 2.5b. The test object was a
gas flame stabilized on a flattened nozzle, and the glass slide
was introduced on the camera side of the second focus. As evident from these records, the fringes can be followed through
regions of large deflections without difficulty. The receptor
was conjugate with the flame at two different magnifications.

Under such circumstances, the relationship be-
tween fringe displacement, ΔY , and angle of deflection, θ,
is:

$$\theta = \frac{\Delta Y(h-f)m}{fh-(l-h)(h-f)} \qquad (2.3)$$

where m is the magnification of the test object while l , f ,
and h represent distances defined in Fig. 2.6, from which Eq.
(2.3) can be easily deduced.

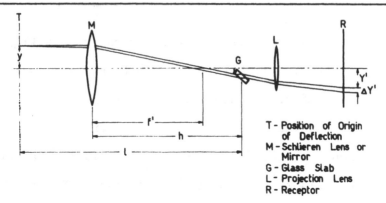

T - Position of Origin
      of Deflection
M - Schlieren Lens or
      Mirror
G - Glass Slab
L - Projection Lens
R - Receptor

Fig. 2.6. Optical arrangement for deflection mapping.

2.4 Shadowgraphy

Shadows are produced, in the absence of a knife-
edge or other marking aperture, as soon as the system is de-
fucused. The sensitivity depends on the amount of defocusing.
This is to be regarded as a facility of the schlieren system
which might occasionally be useful - much simpler equipment is,
of course required for shadowgraphy.

The particular consequence of the properties of
laser light sources is again the breakdown of the geometrical theory at large sensitivity. The theory giving the relevant criteria can
be found in Ref. [12] on pp. 152ff, where the effect of decreasing the
source size has been illustrated (see e. g. plate 21 of Ref. 1.2 re-
produced from Ref. [16] some time before the advent of lasers).

Figure 2.7 (*) illustrates the type of record obtained with a laser source. The superposition of "rays" deflected

(*) See plates pp. 111 ff.

through various paths produces fringes, while the geometrical
theory predicts correctly only the outer envelope of the illumination profile. Because of the difficulty of interpretation
the most useful aspect of such patterns may be the use of their
disappearance as a sensitive criterion for focusing the test
space on the receptor, prior to using schlieren or interfero-
metric methods.

2.5 Interferometry

An interferometer which is suitable for associa-
tion with a schlieren system and its theory have been known
previously [11] , [12] , [27] , [28] . The technique is based
on the use of two small pieces of diffraction grating situated
at the foci of the two schlieren mirrors, or lenses, and so
arranged that adjacent orders from the first grating fill the
two halves of the field of view. One half of the parallel beam
is then used to traverse the test space, while the other acts
as the reference beam. Recombination takes place in the various
orders of the second grating, and one suitable overlapping set
is selected for the camera positon. In particular this has provided a suitable instrument for optical studies of counterflow
diffusion flames [29] .

In the case of laser sources the use of gratings
is both unsuitable and unnecessary. It is unsuitable because

of the small extent of light one obtains in this case at the
focus, which implies diffraction at only one or two of the rul-
ings of the coarse gratings which are required for this purpose,
and because it damages the grating at each flash. It is unnecessary since the coherency of the light makes interferometry so
easy.

      This principle of the simple device which exploits
the monochromaticity of the laser source is illustrated in Fig.
2.8. The schlieren system up to the point of the second mirror
is unchanged, except that only part of the beam is utilized for
the test space, the rest being left unobstructed for use as the
reference beam. In order to superimpose one part of the beam
upon the other, use is made of reflection at the two parallel
surfaces of a glass slab. In contrast to the purpose of the glass
plate of Fig. 2.6, a paral-
lel beam and only one re-
flection at each surface
is now required. Using the parallel beam at the test section would re-

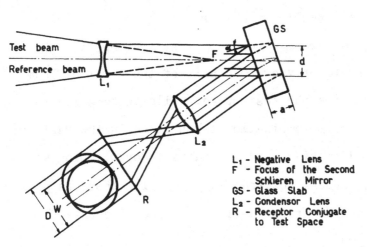

Fig. 2.8. Method of superposition for interferometry.

L₁ – Negative Lens
F – Focus of the Second Schlieren Mirror
GS – Glass Slab
L₂ – Condensor Lens
R – Receptor Conjugate to Test Space

quire an unnecessarily large slab, and the beam is therefore
made parallel again after contraction, by placing a concave lens
whose focus is coincident with that of the second mirror.

The mechanism of superposition is obvious from
Fig. 2.8. As can easily be deduced there, the relation between
the beam diameter, d , thickness of the plate, a , angle of
incidence, α , and extent of overlap, n , in terms of a fraction
of the beam diameter is given by

$$n = \frac{W}{D} = 1 - 2\, \frac{a}{d} \sin \alpha \ . \qquad (2.4)$$

The shorter dimension of the greatest working space possible is
half the mirror diameter. When this is to be achieved $n = 1/2$
and $\frac{a}{d}\sin\alpha = 1/4$. The special case when this is arranged for
$\alpha = 45°$ tends (in the absence of partial silvering) to make
the illumination in the two halves unequal, and diminishes fringe
legibility. The clarity of fringes is improved at more perpendicular incidence. Alternatively one surface may be partially silver
ed so as to make the two illuminations equal at any desired incidence. However, if the schlieren mirrors are sufficiently large
in relation to the test object, as it was indeed the case here,
the required angle of incidence can be obtained without resort-
ing to such optimization.

In the limit of increasing overlap, the beams
superpose exactly with point-to-point correspondence between the
two beams, and the system reverts to deflection mapping. In the

previous section this was used with the transmitted beams beyond the slab. The divergence was exploited there to magnify the sharp, closely spaced fringes obtained at oblique incidence.

If the optics were perfect and used with a parallel-faced slab, the interferometer would produce an infinite fringe adjustement automatically. This is not quite attainable with a mirror-schlieren system, principally because of off-axis aberrations, as illustrated by the photograph of Fig. 2.9 (*) devoid of any perturbation in the test space. The deviation across the 46 cm field is 8 wavelengths at its greatest. When a large test object is examined using such mirror systems, it may be preferable to work with parallel frings. Figures 2.10a (*) and 2.10b (*) are interferograms of gas flames taken near the infinite fringe condition and at a finite fringe spacing respectively (the inclination of the two wave fronts in the latter case was approximately 2.5 wavelengths/cm).

It must be emphasized that this form of interferometer is suitable for use only in conjunction with a laser light source. The reason for this is the fact that while one beam traverses the glass slab twice, the other is reflected on its surface, and the path difference between the two interfering waves is very large indeed, even though the fronts are parallel. The order of fringes recorded in Fig. 2.10 has been, in fact, as large as 10^5.

(*) See plates pp. 111 ff.

2.6 Variable Shear Technique (*)

Of particular significance to the use of laser
light sources for photographic studies of refractive index
fields is the recently developed variable shear interferometry
due to Weinberg and his associates [30] , [31] . The reason for
this is the fact that, similarly as holography, the success of
this technique depends essentially on the coherence length of
the light beam which can be obtained only from a laser.

As all the other methods considered in this
chapter, the optical system is associated in this case as well
with the use of conventional schlieren apparatus. The basic
features of the " variable shear" system are described schematically by Fig. 2.11.
The test section is
situated at the usu
al place in the
schlieren system.
By the use of an ob
jective lens one forms

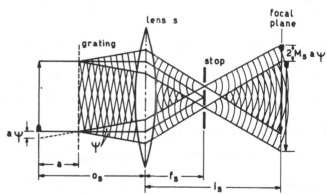

Fig. 2.11. Principle features of variable shear interferometry.

(*) This section has been prepared in collaboration with Dr. J.
W. Meyer to whom the authors are especially obligated for
all the results presented here and many helpful discussions
and explanations.

first a real image of the test object in parallel beam. The
latter is not a necessary condition, but is is pratically convenient and, for this reason, it has been assumed to exist in
Fig. 2.11. The light beam is then passed through a diffraction
grating and focused, by means of a lens, upon a stop. Here the
direct light is blocked off, while only the two first order dif
fraction beams are let through suitable apertures. At the focal
plane of the system one obtains then a region where the two
diffracted beams overlap producing an interference pattern.

As illustrated in Fig. 2.11, immediately after
the grating the wave fronts are inclined to the normal of the
optical axis at the diffraction angle

(2.5) $$\psi = \frac{\lambda}{l} ,$$

where λ is the wave length of the laser light and l the grating line spacing. If the incident beam is undisturbed, one obtains then in the region of overlap a parallel fringe pattern
at a local spacing of $\lambda / 2\psi$, since there are two fringes
formed per wavelength. The fringe spacing in the corresponding
region at the focal plane is then

(2.6) $$\delta = M_s \frac{\lambda}{2\psi} ,$$

where $M_s = i_s / o_s$ is the magnification associated with
lens S .

As it is readily apparent from Fig. 2.12, the

amount of displacement between the two interfering beams at
the focal plane, or the so-called " shear", is

$$s = 2 \, M_s \psi \, a \, , \hspace{3cm} (2.7)$$

where **a** is the distance between the grating and the real image
of the test object. From the above one can see immediately that,
with a given set of optical components, the shear can be modified by a simple expedient of changing the distance **a** , rendering thus the " variable shear" property to the system. By
shifting just the position of the grating one may thus get at
one extreme, using a very small magnitude of **s** , a " differ-
ential" interferometer where fringe shift is proportional to
gradient of the refractive index in the direction of beam displacement, and at the other extreme, with a sufficiently large
s , one may attain interference with a virtually undisturbed
beam, provided only that the region of disturbance is confined
to just one side
of the test sec-
tion which, in
fact, is usually
the case.
For a laser light
with wavelength
of an order of
0.65 microns

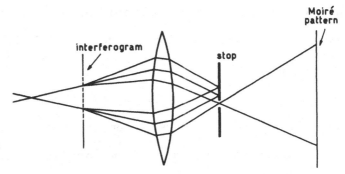

Fig. 2.12. Optical arrangement for reconstruction.

(6943 Å for ruby and 6328 Å for helium–neon), the grating for
the variable shear interferometry should have a spacing of a-
bout 40 lines/mm. According to Eqs. (2.5) and (2.6), with a
normal magnification of $M_s = 2$, this is also the fringe spacing at the focal plane – much too small to be seen by eye, but
sufficiently coarse to be recorded by a Type 55 P/N Polaroid
film whose resolution is of an order of 200 lines/mm, in contrast to holographic emulsions, such as the Agfa–Gavaert Scien-
tia 10E75, whose resolution is not less than 2000 lines/mm. The
photographic record one thus obtains is of "fine fringe" structure.

In order to attain properties of conventional
interferometry, one uses thus a double exposure technique. The
interference fringes are thereby obtained in the form of a Moiré
pattern between the two records. By obtaining the first exposure before the event and the other during the event to be recorded, one gains one more valuable advantage in having all the
effects due to imperfections of the optical system automatically cancelled out.

In order to observe the Moiré pattern thus obtained, the record has to be reconstructed. For this purpose it
must be made on a trasparent film (or plate) and then illumina-
ted by a sufficiently well collimated light beam so that it
could be well focused after being transmitted through the "fine
fringe" record plate, as shown in Fig. 2.12. At focus one places

then a stop with a single aperture that lets through only the light of just one of the first order diffraction beams. Since in effect, the Moiré pattern obliterates the "fine fringe" structure, at the focal plane of the reconstruction system one obtains thus dark areas, while the rest is bleached out by the diffracted light.

Now, a finite fringe spacing of a conventional interferograph is obtained either by rotating the gratings by a small angle between the two exposures, or by recording each exposure on a separate transparency to be then superimposed at an arbitrary angle in the reconstruction process. If the grating is not rotated one obtains an infinite fringe record where dark lines are loci of constant optical paths,

$$P = \int_0^X \frac{dx}{\lambda_t} - \int_0^X \frac{dx}{\lambda_r} \, , \qquad (2.8a)$$

or, since the light frequency is invariant,

$$P = \frac{1}{\lambda_v} \int_0^X (n_t - n_r) \, dx \, , \qquad (2.8b)$$

where x is the space coordinate along the optical axis, X the depth of the test section, n – the refractive index, subscripts t and r denote the test and the reference exposure, respectively, while v stands for vacuum.

Figure 2.13 (*) is a photograph of an infinite fringe, absolute (or unperturbed beam) interferogram of a flame, Fig. 2.14 (*) is a finite fringe, absolute interferogram of the same

---

(*) See plates pp. 111 ff.

flame. Both records were obtained by the use of a helium–neon
laser, the latter corresponding to a rotation of a grating of
40 lines/mm by an angle of approximately 0.01 radian in a system with magnification $M_s = 2$.

With the use of a solid state laser, the coherence length is too small to produce an absolute interferogram.
One has to be content then with a small shear record. This is
illustrated by Fig. 2.15 (*) representing such a record of a finite
fringe interferogram of a helium jet confined between two glass
plates, approximately 6mm apart, in air at atmospheric pressure.

If the amount of shear, s , is in this case known
(it can be easily measured from the lateral shift produced on the
record by a sharp object in the test section) such a record can be
graphically reconstructed to yield a plot of the effective path
length at any lateral position in the test section. Since the recorded optical path at any lateral position y , the space coordinate in
the direction normal to optical axis, is then

(2.9) $$P(y) = Q(y) - Q(y - s) ,$$

one has for this purpose to determine only the component $Q(y)$
The graphical reconstruction procedure based on this principle
is demonstrated in Fig. 2.16. Starting from the point where
the disturbance first starts to manifest itself one has, over
the first strip of width s , $Q(y - s) = 0$. Thus here $Q_1(y) = P(y)$.

_____

(*) See plates pp. 111 ff.

while in the second strip

$$Q_2(y - s) = Q_1(y) .$$

The corresponding $Q_2(y)$ is then obtained by the addition specified by Eq. (2.9). By continuing this procedure strip by strip, one can determine finally both $Q(y)$ and $Q(y - s)$ over the whole extent of the lateral distance from the optical axis, as shown in Fig. 2.16.

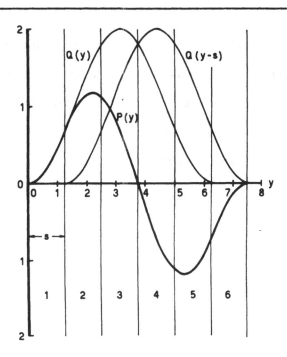

Fig. 2.16. Graphical reconstruction of an absolute interferogram from a small shear record.

In the case of relatively slowly varying refractive index fields, such as flames, a small 1 milliwatt helium-neon laser is quite sufficient for photographing a " variable shear" interferogram using for this purpose a Polaroid film Type 55P/N. Thus the method offers quite an inexpensive and versatile means for the study of such physical processes.

For laser cinematography of explosions, however, some additional problems arise due primarily to the demand for very short exposure times that impose a requirement for light pulses of extremely high intensity. In principle this can be attained by a laser amplifier system combined with the use of proper film emulsion, but in practice such an apparatus would be, of course,

quite expensive. There is also a secondary problem associated
with the necessity for an exact superposition of two records
obtained with the double exposure technique for reconstruction,
but this, we believe, ca be solved by the development of a proper alignment technique.

References

[1] Stroke, G.W., <u>An introduction to Coherent Optics and Holo-graphy</u>, 270pp., Academic Press, New-York, 1966.

[2] Gabor, D., "Microscopy by Reconstructed Wave-Fronts", <u>Proc. Roy. Soc.</u>, <u>A197</u>, 454-487, 1949.

[3] Weinberg, F.J., "A Versatile Apparatus for the Study of Refractive Index Fields in Gases", ARL 63-45, Wright-Patterson AFB, Ohio, February 1963.

[4] Oppenheim, A.K., Urtiew, P.A. and Weinberg, F.J., "On the Use of Laser Light Sources in Schlieren-Interfero-meter Systems.", <u>Proc. Roy. Soc.</u>, <u>A291</u>, 279-290, 1966.

[5] Heflinger, L.O., Wuerker, R.F. and Brooks, R.E., "Holograph ic Interferometry", <u>J. Appl. Phys.</u>, <u>37</u>, 2, 642-649, February 1966.

[6] Brooks, R. E., Heflinger, L.O. and Wuerker, R.F., "9A9--Pulsed Laser Holograms", <u>IEEE J. of Quantum Elect.</u> <u>QE-2</u>, <u>8</u>, 275-279, August 1966.

[7] Tanner, L.H., "Some Laser Interferometers for Use in Fluid Mechanics", <u>J. Sci. Instru.</u>, <u>42</u>, 834-837, December 1965.

[8] Tanner, L.H., "Some Applications of Holography in Fluid Mechanics", <u>J. Sci. Instru.</u>, <u>43</u>, 81-83, February 1966.

[9] Tanner, L.H., "The Application of Lasers to Time-Resolved Flow Visualization", <u>J. Sci. Instru.</u>, <u>43</u>, 353-358, June 1966.

[10] Tanner, L.H., "The Design of Laser Interferometers for Use in Fluid Mechanics", <u>J. Sci. Instru.</u>, <u>43</u>, 878-

886, December 1966.

[11] Gates, J.W.C., "Holography with Scatter Plates", <u>J. Sci.</u>
      <u>Instru.</u>, Series 2, <u>1</u>, 2, 1968.

[12] Weinberg, F.J., <u>Optics of Flames</u>, 251 pp. (Butterworths,
      Washington, D.C.) 1963.

[13] Shafer, H.J., "Physical Optics Analysis of Image Quality
      in Schlieren Photography", <u>J. Soc. Motion Picture</u>
      <u>Engrs.</u>, <u>53</u>, p. 524, 1949.

[14] Lord Rayleigh, "On Methods for Detecting Small Optical
      Retardations and on the Theory of Foucaults Test",
      <u>Sci. Pap.</u>, Cambridge University Press, <u>6</u>, p. 455,
      1920.

[15] Speak, G.S. and Walters, D.J., "Optical Considerations and
      Limitations of the Schlieren Method", <u>Aeronautic-</u>
      <u>al Research Council, Rpt. & Memo</u>; 2859, 1954.

[16] Linfoot, E.H., "A Contribution to the Theory of the
      Foucault Test", <u>Proc. Roy. Soc.</u>, <u>A186</u>, p. 72,
      1946.

[17] Schardin, H., "Die Schlierenverfahren und Ihre Anwendungen",
      <u>Ergebn. Exakt. Naturw.</u>, 20, p. 303, 1942.

[18] Ellis, O.C. de C. and Morgan, E.

      (a) "The Vibratory Movement in Flames", <u>Trans.</u>
      <u>Faraday Soc.</u> <u>28</u>, p. 826, 1932.

      (b) "The Temperature Gradient in Flames", <u>Trans.</u>
      <u>Faraday Soc.</u>, <u>30</u>, p. 287, 1934.

[19] Burgoyne, J.H. and Weinberg, F.J., "Studies of the Mech-
      anism of Flame Propagation in Premixed Gases",
      <u>Z. Elektrochem.</u> (Bunsengesellschaft Symposium)
      <u>61</u>, p.565, 1967.

[20] Reck, J., Sumi, K. and Weinberg, F.J., "An Optical Method
      of Flame Temperature Measurement, II - Sensitivi<u>

ty and Application", <u>Fuel</u>, <u>35</u>, p. 364, 1956.

[21]   Ronchi, V., <u>La Prova dei Sistemi Ottici</u>, Zanichelli,
       Bologna, 1925.

[22]   Levy, A. and Weinberg, F.J., "Optical Flame Structure
       Studies: Examination of Reaction Rate Laws in
       Lean Ethylene Air Flames", <u>Combustion and Flame</u>,
       <u>3</u>, p. 229, 1959.

[23]   Levy, A. and Weinberg, F.J., "Optical Flame Structure
       Studies: Some Conclusions Concerning the Pro-
       pagation of Flat Flames", <u>VII Symposium (Inter-</u>
       <u>national) on Combustion</u>, p. 296, Butterworths,
       London, 1959.

[24]   Wolter, H., "Schlieren-Phasenkontrast-und Lichtschnit-
       tverfahren", <u>Handb. Phys.</u>, 24, p. 555, 1956.

[25]   Pandya, T.P. and Weinberg, F.J., "The Study of the
       Structure of Laminar Diffusion Flames by Optic-
       al Methods", <u>IX Symposium (International) on</u>
       <u>Combustion</u>, p. 587, Academic Press, New York,
       1963.

[26]   Hannes, H., "The Properties of Shadowgraphs", <u>Optik</u>,
       <u>Stuttgart</u>, 13, p. 34, 1956.

[27]   Kraushaar, R., "A Diffraction Grating Interferometer"
       <u>J. Opt. Soc. Amer.</u>, <u>40</u>, p. 480, 1950.

[28]   Sterrett, J.R. and Erwin, J.R., " Investigation of Dif-
       Fraction Grating Interferometer for Use in Aero-
       dynamic Research" , <u>NACA Tech. Note 2827</u>, 1952

[29]   Pandya, T.P. and Weinberg, F.J., " The Structure of
       Flat Counter-Flow Diffusion Flames" , <u>Proc. Roy</u>.

<u>Soc.</u>, <u>A279</u>, p. 544, 1964.

[30] Schwar, M.J.R., and Weinberg, F.J., " Coherent Light
      Sources and Refractive Index Fields", <u>Phys. Bull.</u>,
      <u>21</u>, 490–492, 1970.

[31] Jones, A.R., Schwar, M.J.R. and Weinberg, F.J., " Gen-
      eralizing Variable Shear Interferometry for the
      Study of Stationary and Moving Refractive Index
      Fields with the Use of Laser Light", <u>Proc. Roy.</u>
      <u>Soc.</u>, <u>A322</u>, 119–135, 1971.

PLATES FOR CHAPTER 2

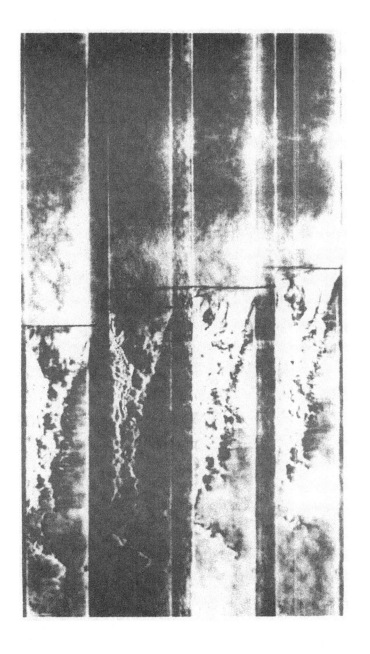

Fig. 2.2. Sequence of schlieren records of the wave processes leading to the development of detonation. Equimolar H_2–O_2 initially at 120 mmHg in rectangular (2.54 x 3.81 cm) tube photographed through shorter side using Kerr-cell modulated ruby-laser at a rate of 2×10^5 frames per second.

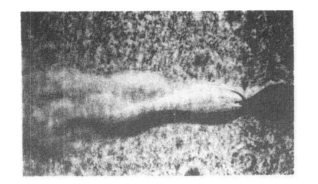

- c -

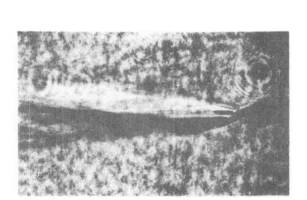

- b -

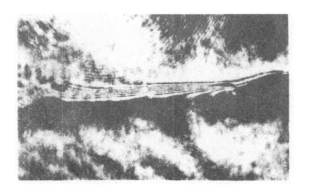

- a -

Fig. 2.3. Schlieren records of a flame obtained with gas laser using marking apertures as follows :

a - knife edge
b - photographically produced neutral wedge
c - quartz prism-polaroid filter combination.

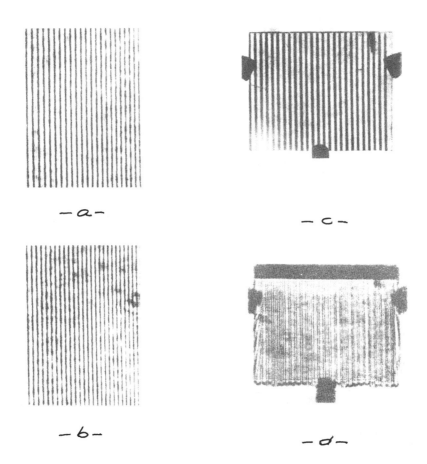

Fig. 2.4. Comparison of legibility of fringes with that of shadows cast by a grid.
    a - fringe marking in the parallel beam at test section produced by inclined glass slab at first focus
    b - fringe marking 2.5 m beyond test section
    c - Ronchi grid located in parallel beam at test section photographed on plate immediately behind it
    d - diffraction spread of the Ronchi grid 2.5 m beyond test section.

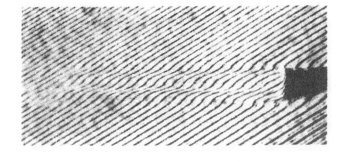

- b -

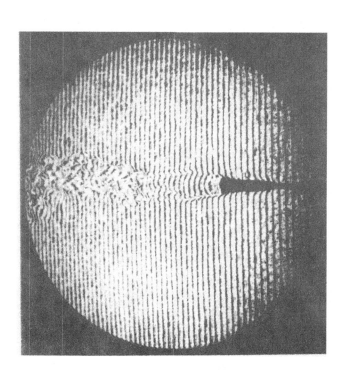

- a -

Fig. 2.5. Records of flame on a slot burner illustrating application of deflection mapping.

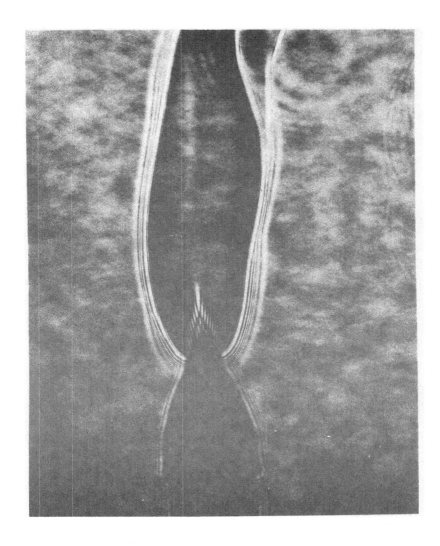

Fig. 2.7. Parallel beam shadow of a flame illustrating diffraction effects obtained by a laser source.

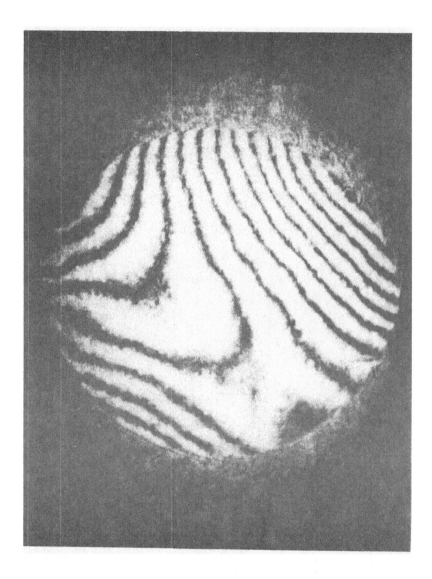

Fig. 2.9. Approach to infinite fringe interferogram of the undisturbed test space (maximum deviation 8 wavelengths over 46 cm).

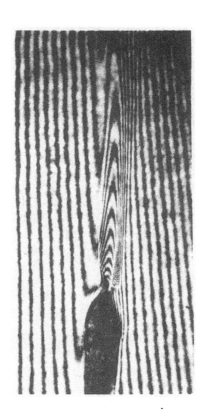

-a-   -b-

Fig. 2.10. Interferograms of flame on a slot burner.
a - near infinite fringe setting
b - at finite fringe setting (2.5 / cm).

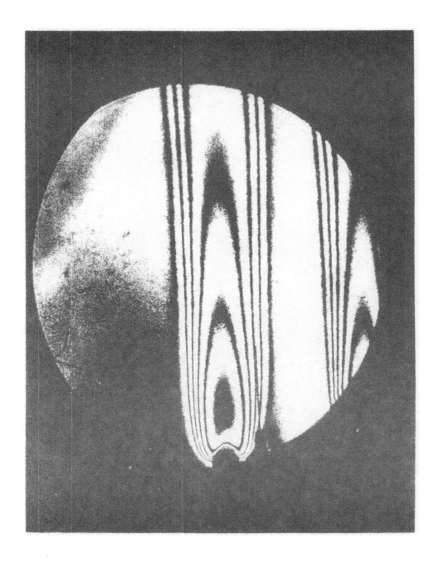

Fig. 2.13. Infinite fringe absolute interferogram of a flame obtained by the variable shear technique.

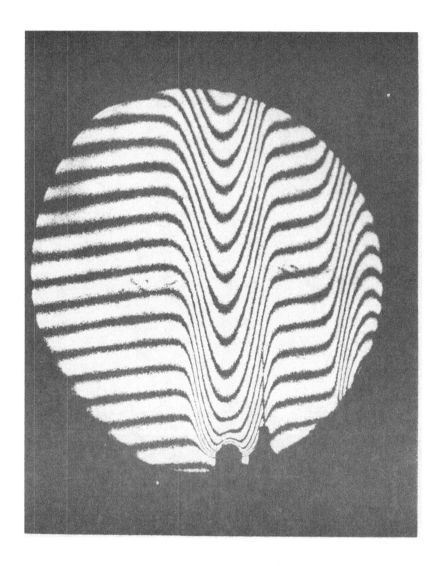

Fig. 2.14. Finite fringe absolute interferogram of a flame obtained by the variable shear technique.

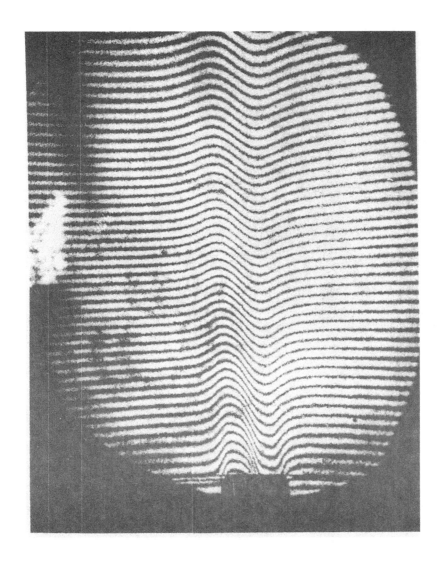

Fig. 2.15. Finite fringe small shear interferogram of a helium jet in air.

CHAPTER 3

ELECTRONICS (*)

3.1 Introduction

High-speed cinematography is a basic research tool of fundamental importance to the study of a vast variety of transient phenomena occurring in many branches of science. For the present purpose, this will be understood to imply framing rates in excess of 10 kHz and exposure times of less than 1 microsecond. In such a regime of operating conditions, high-speed framing cameras, although available, become extremely cumbersome, have limited application, and are expensive. A simpler approach, which utilizes a pulsed light source to determine both exposure duration and framing rate, is generally preferable. In this case, the camera serves only as a means of displacing successive images over the film at a rate sufficient to avoid image overlap.

High-intensity electric spark discharges have

(*) The authors wish to reiterate their appreciation to Dr. G. J. Hecht for the conceptual direction in the development and primary responsability in the design of the apparatus described in this chapter, and to Mr. K. Hom who was mainly responsible for its construction, operation, and maintenance.

been employed successfully for this purpose, but are subject
to several important and severe restrictions. Chief among these
are limited intensity, source size, spectral quality, maximum
repetition rate, and spark duration. These factors are inter-
related; for example, after a spark plasma temperature of a-
bout 10^4 °K is reached, total intensity can be increased only
by enlarging the spark size. A large area source is perfectly
suitable for general illumination if direct real imaging of
real objects is desired, but, if the aim is to record, for example, density gradients (Schlieren photography), then the optical system is capable of accepting only a very limited-area
source. Spectral quality requirements impose an additional severe limitation if the purpose is to record density (optical
path length), as in interferometry, in which the source
of illumination must usually by highly monochromatic. Repetition rate is fundamentally limited by de-ionization times in
gaseous discharges, while spark duration (exposure time) is
difficult to reduce because of stray inductance or capacitance
in the discharge circuitry. Various ingenious devices and design compromises have, in part, overcome some of these limitation, but, in general, it becomes extremely difficult to a-
chieve spark operation with the simultaneous combination of ex
posure times less than 10^{-7} sec, repetition rates greater than
10 kHz, and an energy in each spark of more than 1 J, with an
acceptable degree of reliability.

The advent of laser technology has fundamentally lifted these restrictions. A pulsed solid-state laser, such as neodymium or ruby, produces a coherent light pulse, which immediately represents an excellent approximation to the optical equivalent of a point source emitting almost monochromatic light with all of the energy concentrated into a beam of narrow divergence. Furthermore, the energy available, for example, in a pulsed laser beam is capable of achieving very high values - a 1J laser is rather modest. The " giant pulse" mode of laser operation, produced by electro-optical effects, was first demonstrated by McClung and Hellwarth, [1](*) by the introduction of a Kerr cell into the laser cavity for Q-spoiling. This increases by several orders of magnitude the output light-pulse power and decreases the effective light-pulse duration down to the order of 10^{-8} sec, with absolute control of pulse timing. Ellis and Fourney [2] further controlled the Kerr cell to yield a series of light pulses whose repetition rate could be varied, thus producing a light source which is ideally suited for many applications in the field of high-speed photography for the study of transient phenomena.

In this chapter, after a brief introduction of the laser principle, electro-optical pulse control techniques, such as Kerr and Pockels cells, will be described. Laser Q-switch-

---

(*) Numbers in square brackets denote references listed at the end of the Chapter.

ing using bleachable absorbers are also considered, followed
by the description of a Kerr-cell controlled laser light source
that is designed specifically for use in schlieren cinematography
of non-steady phenomena in explosive gas mixtures. This type of
laser source, however, can be easily adapted to other optical
systems.

3.2 Laser Principle

The principle was first proposed by Schawlow
and Townes [3] in 1958, and in 1960 Maiman [4] demonstrated the
first practical laser operating in the visible spectrum. This
first laser, the pulsed ruby system, remains one of the most
useful of the many types now available.

A solid state laser, such as ruby or neodymium,
consists of a fluorescent solid composed of a small amount of
an active ion (Cr^{+3} or Nd^{+3}) doped into a transparent host
material (alumina crystal or glass). According to quantum
theory, valence electrons of the active ions are restrict-
ed to discrete values of energy which can be modified to
a certain degree by the host material. The lowest energy
allowable is the ground state, and any larger permitted
energy puts the electron to a higher quantum level from
which it subsequently returns to the ground state by radiation emitting a photon of light with a frequency dependent

on the energy lost in transition. In a good laser material,
transition between a dominant excited state and the ground
state account for almost all of the light radiation.
With the additional effects of line—narrowing, the laser
produces highly monochromatic light.

Stimulated photon emission occurs when a previously emitted photon, traveling through the laser material,
causes an excited electron to emit a new photon. By this
process, the laser (Light Amplification by Stimulated Emission of Radiation) acts as a light amplifier. The fact that
this light is radiated in phase and direction with the
stimulating photons accounts for the highly coherent nature
of the laser beam.

The operation of solid state lasers can be described as follows. Let N_0 denote the number density of atoms,
or population, of the ground level, and let N_1 denote the popu
lation of the dominant excited state in a laser—material rod
of length l . The amount of amplification, or gain, a light wave
receives through stimulated emission in passing through the rod
depends on the number density of excited atoms N_1 and is given
by the following relation:

$$G_E = e^{N_1 \sigma l} \, , \tag{3.1}$$

where σ is the absorption cross-section.

Light making one pass through the rod is also

absorbed by ground-state electrons and the corresponding attenuation, depending on N_0, is

(3.2) $$G_A = e^{-N_0 \sigma l}.$$

In a real material there is another source of
light attenuation, due to optical imperfections (such as impu-
rities, strains, or bubbles) in the laser rod. Such effects can
be characterized by an absorption coefficient α and the resulting attenuation is

(3.3) $$G_R = e^{-\alpha l}.$$

In a practical system, there is typically a 100%
reflecting mirror at one end of the rod, and a mirror with reflectivity r at the other end. Therefore, light making one pass
in each direction through the rod experiences a gain of

(3.4a) $$G_T = r(G_E G_A G_R)^2$$

or

(3.4b) $$G_T = e^{2\sigma(N_1 - N_0) - 2\alpha l + \ln r}.$$

The total gain will be greater than unity if

(3.5) $$N_1 - N_0 > \frac{\alpha}{\sigma} - \frac{\ln r}{2\sigma l}.$$

Equation (3.5) is known as the threshold condition and specifies the minimum difference between the popula-

tions of the two states required to achieve laser action. When
N_1 exceeds N_0 , the population is said to be _inverted_ and
a condition of _negative temperature_ is achieved.

When the laser rod is in thermal equilibrium
(no external energy applied), N_1 is related to N_0 by the
Boltzmann factor

$$\left(\frac{N_1}{N_0}\right)_{equilibrium} = e^{-E_{01}/kT} \qquad (3.6)$$

where E_{01} , the energy difference between the two states, has
the value of approximately 2 eV for visible radiation,
while kT , Boltzmann's constant multiplied by the absolute
temperature of the rod, is about 0.025 eV at room temperature. This gives the value of 10^{-35} for the Boltzmann factor
so that, for all practical purposes, the entire population may
be considered as initially in the ground state.

Therefore, to get lasing action, external ener-
gy must be applied to the laser rod, increasing N_1 , until Eq.
(3.5) is satisfied. In the solid-state laser, this is achieved
by subjecting the rod to an intense white pumping light which
excites electrons into the levels of band E_2 , as shown in Fig.
3.1 for the case of pink ruby. These ions rapidly drop to level
E_1 by nonradiative transitions (exciting molecular and lattice
vibrations). Since the E_1 level is long-lived (5×10^{-3} sec),
the net effect is to populate the dominant excited state. When

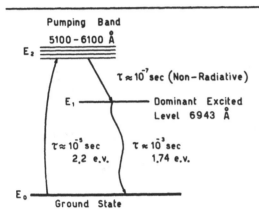

Fig. 3.1. Energy-level diagram for pink ruby.

the threshold condition of Eq. (3.5) is satisfied, the system is capable of lasing action. The mirrors at each end of the laser produce a high Q cavity, i.e. any photon ray path not parallel to the rod axis (perpendicular to the mirror faces) rapidly "walks off" the mirrors and is lost before much amplification has taken place. Rays that travel along the axis, on the other hand, may make many reflections within the cavity and be amplified again on each successive transit. Therefore the light that passes through the transmitting mirror (the useful output end) will not only, as discussed before, be nearly monochromatic and coherent, but also very intense and nearly parallel.

A useful property of many solid-state laser materials is their inherent dichroism. This is a property of certain crystals, such as the pink ruby, that produces preferential transmission of linearly polarized light (normally perpendicular to the optic axis). It is this property then that is exploited by means of the electro-optical shutter technique described in the next section to produce the "giant pulse" or the stroboscopic effect required for laser cinematography.

3.3 Principle of Electro-Optical Pulse Control

The pulsating emission pattern of solid state lasers is most irregular and thus becomes a disturbing factor in many practical applications, especially in cinematography, where the control of timing and light intensity is essential. This difficulty may be overcome by regulating the regenerative cycle of the laser, a method first proposed by Hellwarth [5] , that, in addition to controlling the pulsating time, enhances the peak light intensity by orders of magnitude. In its simplest form this method involves the introduction of a fast shutter or a chopper wheel, such as that used by Collins and Kisliuk [6] inside the optical cavity. With the shutter closed, the population inversion in the ruby crystal, produced by the flash lamps, reaches a magnitude far beyond the threshold level for normal lasing. When the shutter opens, a rapid build up of excited molecules takes place and the laser discharges its radiation in an extremely short pulse having very high light intensity which, for obvious reasons, is called a "giant pulse".

The method described above is but one of various techniques that utilize the obstruction of the optical path in a laser cavity in order to delay, or control, the onset of laser oscillation. This process is referred to as Q-switching, or Q-spoiling, of the cavity, a term having its origin in radio engineering, since the laser cavity itself may be looked upon as an optical resonator analogous to an oscillating electronic circuit.

Since mechanical shutters have an inherent speed limitation, and since installing them inside the laser cavity is, to say the least, inconvenient, the most popular today is the Q-switching attained by means of electro-optical effects that utilize the application of an electric field to open or close an optical shutter [1]. Their operating principle, as well as that of the magneto-optical devices, is based directly on the electromagnetic nature of light, manifesting the interaction between matter and light when the former is subjected to an external electric or magnetic field. This interaction causes some transparent materials to exhibit birefringence (or double refraction, see Appendix A) when placed in an electric field by causing the material to behave like an uni-axial doubly refracting crystal with its optic axis oriented in the direction of the applied field. This phenomenon is observed in gases, liquids and solids. In the latter, however, it is more complicated because the electric field produces electro-striction which in turn produces stress birefringence.

The best known electro-optical shutter is the Kerr-cell [7 - 9]. It consists of a rectangular glass cell containing a liquid that exhibits birefringence under the influence of electric fields applied by means of two parallel plate elec-

trodes enclosed
within the cell.
Figure 3.2 de-
picts such a
cell placed be-
tween cross po-
larizer and analyzer elements.

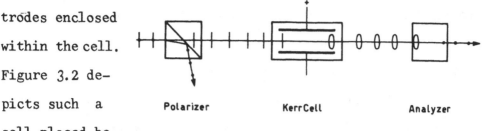

Polarizer   KerrCell   Analyzer

Fig. 3.2. A schematic of a Kerr-cell electro-optical shutter.

The action of the Kerr effect in a birefringent
medium can be described simply as follows. The difference between the refractive indices of the E-ray and the O-ray produced by the cell can be expressed as:

$$n_E - n_0 = B \lambda E^2 \qquad (3.7)$$

where B is the Kerr constant, λ the light wave length, and E
the amplitude of the electric field vector. Elementary properties of birefringence are described in Appendix A. Using Eq.
(A.3) given there, Eq. (3.7) yields the following expression
for the phase difference between the two components:

$$\delta = 2 \pi l B E^2 . \qquad (3.8)$$

Thus it is seen that, for a given cell, any degree of ellipticity may be produced from plane polarized light by varying E.

The values of the Kerr constant B vary great-
ly from substance to substance as shown in Table 1. Since nitrobenzene has a markedly higher value of B, it requires, accord-

ing to Eq. (3.8), lower electric field intensity than any other medium, to produce the desired effect and, hence, it is the most commonly used substance for Kerr cells.

For normal operation as a shutter, with reference to Fig. A.2 (*), the cell has its electrodes tipped at 45° to the polarization plane of the plane polarized light beam without any voltage between the electrodes, the medium inside the cell is isotropic and the analyzer blocks the light completely. Under such no voltage condition, $n_E = n_0$ so that both components of the incident wave A propagate with the same velocity and no ellipticity is produced.

With the voltage applied, $n_E > n_0$, so that the 0 - ray falls behind and out of phase with the E - ray. The more n_E exceeds n_0 , the greater becomes the phase difference, so that various degrees of ellipticity are produced. Since the analyzer will pass any of the horizontal component, the combination of the Kerr cell with the crossed polarizer-analyzer system constitutes a shutter which is electrically operated. The fractional loss in the light intensity, I , transmitted through such a shutter, is [10] :

$$(3.9) \qquad\qquad \frac{I}{I_0} = \sin^2(\pi\, B\, l\, E^2)$$

where I_0 is the incident intensity.

In the laser cavity, as shown in Fig. (3.3) (*), there is only a single polarizing element acting as both the polarizer

(*) See p. 188 and p. 151

and analyzer while in each cycle of oscillations the ray passes twice through it, being reflected by the back mirror which inverts the sense of elliptical polarization and yields a phase shift of $\delta = \frac{\pi}{2}$ (see Fig. A.3 (*)).

Thus, in order to attain the phase shift required for the operation of the shutter, one needs δ of $45°$. Equation (3.8) gives then for the corresponding voltage

$$V = Ed = \frac{d}{\sqrt{8Bl}} . \qquad (3.10)$$

The capacitance of the plates, on the other hand, is given by

$$C = \frac{k}{0.9 \times 4\pi} \frac{lw}{d} \quad \text{microfarads} \qquad (3.11)$$

where k is the dielectric constant, and w is the width of the electrodes.

In our particular application, $d = 1/4''$ and $w = 0.625$. To obtain $C \approx 15\mu fd$, while $k = 35$ for nitrobenzene at room temperature, Eq. (3.13) yields $l = 2cm$. Then Eq. (3.13) the required voltage of approximately $10\,KV$. Finally for a given driving current i, one may evaluate the rise time from the relation

$$i = C\frac{dV}{dt} \qquad (3.12)$$

assuming a ramp-shaped voltage pulse. For $i = 8\,amp.$, this yields a rise time of 15 nanoseconds.

Another electro-optical device which found wide

(*) See p. 189

application as a laser-pulse control device is the Pockels effect [11] , [12] which utilizes a certain class of crystals which, under the influence of an electric field produce displacements of the particles in its lattice that affect its refracted properties. This shutter has several advantages over the Kerr cell. Since the effect is larger, the applied voltage can be lower; it uses a solid crystal instead of the highly toxic liquid, nitrobenzene; and the effect is linear with voltage, [13] whereas the Kerr effect is proportional to the square of the voltage. Many crystals of the form XH_2PO_4 show a sizeable Pockels effect. One such crystal is potassium dihydrogen phosphate (KH_2PO_4), commonly called KDP , another, is the ammonium dihydrogen phosphate ($NH_3H_2PO_4$), or ADP .

From our own experience, however, it was found that, while Pockels cells are very useful in the production of single laser giant pulses, their surfaces tend to burn under the thermal effects in high frequency stroboscopic operations. For this reason, we shall not dwell here on their theory of operation, while the interested reader is referred to Ref. [13] .

3.4 Principle of Passive Pulse Control

The electro-optical pulse control devices described in the previous section are well suited for the cinemato-

graphic applications. However, as will be seen in the next section, they require elaborate electronic circuitry for their operation and control which puts them within reach of only well
equipped laboratories. A passive pulse control device, on the
other hand, is based on the optical excitation of some special
molecular species that are capable of absorbing radiation energy
at laser frequency to a definite level. Up to that level they
act therefore as an efficient absorber, while beyond it they
become virtually transparent, thus simulating a closed shutter
and an open one, respectively. This method does not involve the
use of electronic elements and is, consequently, much easier to
use. Even though such devices are exploited here only to produce single giant pulses, such as those used for ignition in
Chapter 1, a brief description of them is included for their
potential use in laser cinematography.

The most commonly used bleachable absorber de-
vice, and by far the simplest, consists of a small cell contain
ing one of several phthalocyanine solutions, that are essential
ly metallo-organic compounds [14] , [15] . By judicious choice,
one may obtain a solution that has its peak absorption frequency coincident with that of the laser. For example, a solution
of chloroaluminium phthalocayne in chloronaphthalene has its
maximum absorption near the ruby laser frequency [15] . Upon
pumping, the absorber prevents the occurrence of any net amplification of light since a much larger ratio of the ions in the rod

(Cr^{+3} or Nd^{+3}) are excited to the higher energy level than that required for normal lasing. As the rate of generation of photons in the rod itself, due to pumping, exceeds the rate of their annihilation, due to absorption, the laser begins to emit weak coherent radiation. Since only a relatively small number of phthalocyanine molecules need be present in the cell, due to their large cross-sections for absorption of the laser light, this weak lasing action is sufficient to saturate the absorption at the laser frequency. Thus, chloroaluminium phthalocyanine gets to be bleached and it becomes virtually transparent. When this happens the photons generated in the meantime by the laser rod are suddenly released producing, upon amplification by interreflections between the mirrors bounding the laser cavity, a "giant pulse". The energy of the pulse thus emitted consumes the existing inverted population and the laser rod becomes "de-excited", terminating the pulse and causing a concomitant de-excitation of the cell. In principle, such a sequel of events can be repeated, although pratically the attainment of a regular frequency, as required for laser cinematography, may present some problems due to the complicated de-excitation process of the phthalocyanine molecules.

Phthalocyanine compounds are by no means the only bleachable absorbers used. Cryptocynine dyes [16] , [17] are used quite successfully in connection with Q -switching of ruby lasers. An example of that is the Kodak's Eastman Q-

Switch Solution A10220 that consists of 1,1 - diethyl - 4,4' - carbocyanine iodide dissolved in a cetonitrile. This solution has its peak absorption at 0.705μ which is sufficiently close to the ruby frequency of 0.694μ, to produce the desired effect.

Another promising technique is that developed by Stark et al. [18] for the production of giant pulses in ruby lasers by the incorporation of a rod of uranyl-glass in the laser cavity and irradiating it with the flash lamps used to pump the rod. Prior to its exposure to the pumping light, the uranyl (UO_2^{++}) ion is completely transparent at the ruby laser frequency. Upon exposure to flash light, however, some excited states develop ($\nu \approx 20,000$ cm^{-1}) that are intensely absorbing at this frequency. These states, in turn, may be bleached in a manner similar to that described above for phthalocyanines to produce a giant pulse. The broad absorption band of these excited states make such a device applicable to neodymuim-glass lasers as well [15] .

A technique analogous to that utilizing uranyl-glass was used by Cross and Cheng [19] in a successful attempt to produce a train of pulses from a neodymium laser at a repetition rate in excess of 1KHz in frequency. For this purpose the cell containing the solution of rose bengal (tetrachloro- tetra-iodo-fluorescein) in isopropyl alcohol was placed in the cavity. The dye was thus optically pumped at the same time as the laser rod. A regular series of pulses resulted with a spacing of ap-

proximately 125 microseconds. While this rate of firing is not
sufficiently high for the study of explosion phenomena, this
technique manifests the potential of passive Q -switch devices
in high-speed laser cinematography.

   Finally, one should note that the use of bleachable dyes for Q -switching improves mode selection [16] due
to the narror absorption band in the dye cell. This results in
a longer coherence length that should be of particular interest
to the "variable shear" technique described in Chapter 2.

3.5 Apparatus

   Having the fundamental principles of the operation and modulation of laser oscillators presented in previous
sections, a particular application of these principles will be
here described with reference to a Kerr-cell controlled ruby
laser designed specifically for laser cinematography.

(a) Laser Head

   Typical laser heads, or cavities, are discribed
schematically by block diagrams in Fig. 3.3.

   The essential element of a head is the lasing
material, a ruby crystal in this case, that is situated in the
cavity bounded, on one side of the optic exis, by a reflecting mirror (an optical flat coated with a dielectric yielding
a reflectivity in excess of 99% at the laser frequency) and,

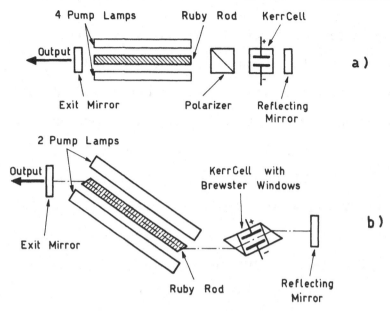

Fig. 3.3. a) A block diagram of straight laser heat
b) A block diagram of Brewster-angle laser head .

on the other side, by the exit mirror (a fused silica etalon
with a reflectivity of 27%).

The laser material is excited, or pumped, by
flash lamps and the oscillator cavity is modulated by Q - spoil
ing obtained by means of a Kerr-cell system.

Figure 3.3 presents two basic systems:

(a) straight, (b) Brewster angle.

In the first, the ruby rod is cut at right angles to the axis
and the windows of the Kerr-cell are parallel to the end faces
of the rod. In this case one needs, for proper operation, an
additional plane-polarizing element in the cavity, such as a
Glan-air space prism or a glass plate of suitable thickness inclined to the optical axis by the Brewster angle [8] , the use

of the latter being, of course, associated with a loss in light intensity. In the second system the ruby rod itself is cut at the Brewster angle, while the Kerr-cell is provided with Brewster angle prism-windows, obviating thus the need for a polarizing element and rendering the cavity maximum efficiency.

Figures 3.4 (*) and 3.5 (*) are photographs of the two cavities. The first has a Verneuil-grown ruby rod 5/8" (15.9 mm) in diameter and uses four flash tubes. The second is equipped with a high quality Czochralski-grown ruby crystal 1/4" (6.45 mm) in diameter and uses two flash tubes. The Czochralski rod has a higher gain and smaller beam divergence, yielding light output of higher intensity and more uniformity than the Verneuil rod. Since in the process of each operating cycle the ruby rod is heated to a temperature which is sufficiently high to prevent its normal function, it has to be an coded and sufficient time allowed (about 5 minutes) for the temperature to drop down to operational level before next experiment is started.

(b) Electronic Circuitry

The large diameter laser system represented by Fig. 3.4, requires a relatively high electric potential for its electronic control is in this case provided by a vacuum tube circuitry. The block diagram of such a system is presented in Fig. 3.6. The system is trigged by a 100VDC pulse which actuates at the same time the pump lamp power supply whose elements are enclosed by a broken line in Fig. 3.6, and the EOLM driver circuit. Since it takes about 500 microseconds for

(*) See plates pp. 177 ff.

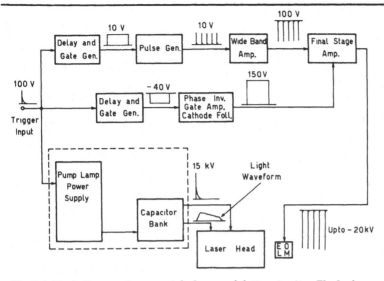

Fig. 3.6. Block diagram of vacuum tube laser-modulation circuitry. The broken line encloses the laser pump lamp power supply.

the N_1 population to be pumped up to meet the thresh-old condition of Eq. (3.5), the EOLM drive is delayed for that amount of time using a delay and gate genera-tor (EH Research Laboratories Inc.) that produces a square signal hav-ing a height of 10 V and 200 micorseconds width. This signal is fed in-to a pulse generator (Computer Measurement Co.)causing it to emit a train of pulses, each for about 10 nanoseconds in half-width, at an exactly controlled frequency by means of a time-mark generator, while retaining the original 10 V height. These signals then enter a wide band amplifier (Hewlett-Packard 460 B) emerging at a 100 V strength, retaining the 10 nanoseconds half-width. They pass then through a final stage high voltage pulse amplifier, Fig. 3.7, based on the use of two high power EIMAC-4PR60 B tetrodes, yielding a series of 75 nanosecond pulses, each at a potential of up to 20 KVDC that last for approximately 200 microseconds to drive the Kerr-cell. In order for the final stage amplifier to operate, a bias of 150 VDC must be imposed on it during modulation. This is accomplish-ed by the delay and gate generator (Digital Equipment Co.) and

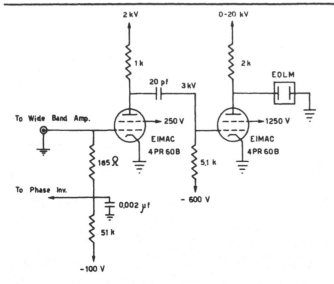

Fig. 3.7. Pulse amplifier circuitry for vacuum tube laser modulator.

the phase inverter, gate amplifier and cathode follower, shown in Fig. 3.8. In order to attain ultra-short light pulses, a smaller diameter laser system, represented by Fig. 3.5, has been developed so that it could be operated at a lower voltage permitting the use of considerably sharper pulses than those obtainable by the use of the vacuum tube system described above. The driver circuitry designed and built for this purpose by C.C. Lo [20] exploits avalanche transistors and toroidal trans-

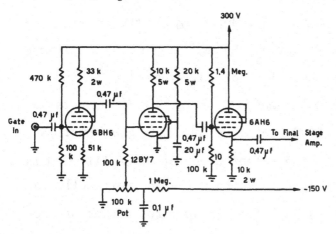

Fig. 3.8. Phase inverter, amplifier, and cathod follower circuitry for vacuum tube laser modulator.

formers. The latter had to be hand wound, and their specifications are given in Table 3.2 (*). Except for the final stage driver sector, the whole system is in this case composed of solid state elements. Its block diagram is given by Fig. 3.9.

The exact time control is provided in this case by a sine-wave generator (Hewlett-Packard 241A) whose output is fed into a shaper that converts the sine-wave into a continuous train of pulses of about 15 V amplitude and 40 nanoseconds in half-width. The pulses enter then a gating circuit which is delayed by about 500 microseconds from the initial trigger pulse, to allow sufficient time for the buildup of N_1 population in the laser rod, producing thereupon an approximately 200 microseconds long train of pulses.

The gated pulse train is then led to the driver unit where it triggers two avalanche transistors that drive a planar triode ML7815. Finally, the driving pulses are fed, through a step-down inverting transformer, to the final stage by a 125 ohm coaxial cable which is terminated in 125 ohms at the grid of an ML8538 planar triode energized to a potential of up to 9 KVDC by an external power supply. The Kerr-cell is connected in parallel with the anode of the triode and is thus maintained normally under its bias voltage. Thus with no external trigger input to the gate delay circuit, the EOLM call is in the closed-shutter position. When it opens the individual pulses are fed to the grid of the triode.

The pump lamp driving unit shown at the bottom of Fig. 3.9 is the same as that presented in Fig. 3.6.

The details of the shaper circuit are described

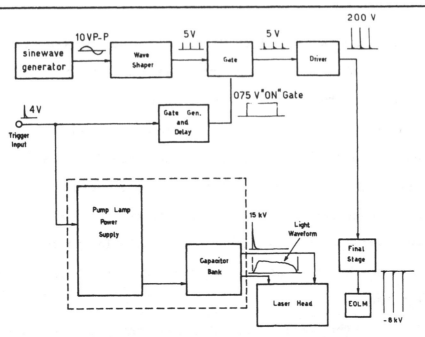

Fig. 3.9. Block diagram of solid state laser modulation circuitry. The broken line encloses the laser pump lamp power supply.

in Fig. 3.10. Transistor networks Q1 and Q2 convert the sine-wave signal into a square wave; Q3 amplifies the square wave before it is differentiated by the RC network C3 and R9, while Q4 serves as an emitter-follower and buffer. The output pulse of this stage has a rise time of about 15 nanoseconds and a base width of 100 nanoseconds. The rise time stays essentially the same for an input frequency range from 200 Hz to 4 MHz, and the peak of output pulse amplitude is 5 to 7 V. The gate delay and forming cir-

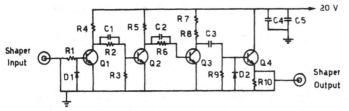

Fig. 3.10. Shaped circuitry for solid state laser modulator. Specifications of the elementary components in this circuit are given in Tables 3.3 – 3.5.

cuit are schown in Fig. 3.11. Tran-

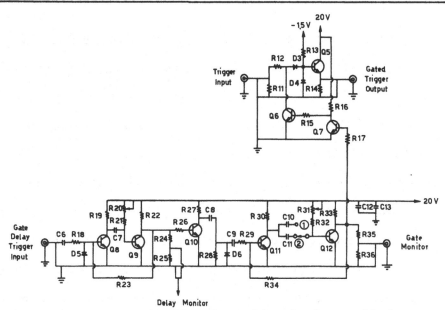

Fig. 3.11. Delay and gate circuitry for solid state laser modulator. Specifications of the elementary components in this circuit are given in Tables 3.6. − 3.8.

sistor networks Q8 and Q9 form a one-shot multivibrator yielding a negative square wave whose pulse width is adjusted by variable resistor R20 to produce a time delay over a range of from 100 microseconds to 1 millisecond. The multivibrator output is amplified by transistor network Q10 and differentiated by the RC network C8 and R28, yielding two sharp pulses of opposite polarity, while diode D6 clips off the negative spike. Transistor networks Q11 and Q12 form another one-shot multivibrator which, by means of variable resistor R31 and switch S1, controls the width of the gate over a time range of from 5 to 500 microseconds.

The gating circuit itself is formed by transistor networks Q7, Q6 and Q5. The transistor Q7 is normally off, while Q6 is normally on, thus acting as the low leg of a

voltage divider whose other leg is resistor R12. Diode D3
further blocks the low level output signal voltage of the divider, while transistor Q5 serves as a buffer. A negative bias
is fed through R13 to the base of Q5 helping it to recover
faster after each time it conducts. This circuit yields an attenuation factor of at least 50 db. The output pulse has a rise
time of about 15 nanoseconds, and an amplitude of 2 V that is
used to trigger the avalanche transistors in the driver stage.

The circuit of the driver stage is given by Fig.
3.12. Transistors Q13 and Q14 are specially selected to operate in the avalanche mode [21] . The type used in this pulser
is Motorola 2N2222 having time delay less than 2 nanoseconds
and rise time less than one nanosecond. Since Q13 and Q14
are connected in series, an isolation transformer is used at
the input. The bias current of Q13 and Q14 is approximately
200 microamperes and the
emitter of Q13 is con-
nected directly to the
grid of the planar triode
V1. When transistors Q13
and Q14 turn on, the ca-
pacitor C15, normally
charged up to 120 V, is
switched right across the
grid of V1, thus turning

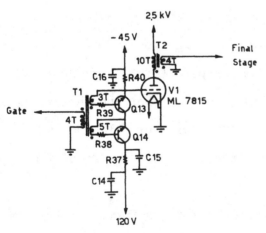

Fig. 3.12. Driver stage circuitry for solid laser modulator.
Specifications of the elementary components in this circuit are given in Tables 3.2 and 3.6 – 3.8.

it on. The transformer T_2 inverts the polarity of the pulse from negative to positive as well as steps down the 2500 V pulse to one of an
amplitude of 450 V , increasing thus the current to the level requir
ed to drive the final stage amplifier to which it is conveyed by an
RG2130 cable. The resulting output pulse has a rise time of 5 nanoseconds and a halfwidth of approximately 25 nanoseconds.

The circuit diagram of the final state is presented in Fig. 3.13. In order to minimize the time constant of
the driver circuitry, this stage has to be situated in the immediate vicinity of the EOLM cell and for this reason it is
mounted in the laser cavity as shown in Fig. 3.5 (*). The primary
element of the final stage is a planar triode, V_2 ,(Machlett

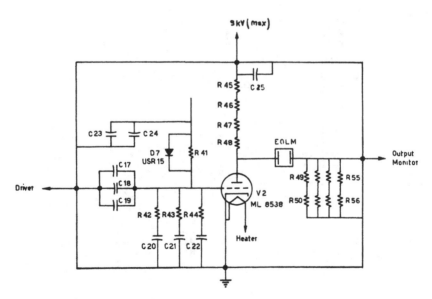

Fig. 3.13. Final stage amplifier circuitry for solid state laser modulator. Specifications of
the elementary components in this circuit are given in Tables 3.9 and 3.10.

(*) See plates pp. 179 ff.

type ML 8538) which is normally biased off by a negative 140
VDC potential applied to its grid. A 1.4 – ohm current shunt
is connected in series with the EOLM cell to serve as an output monitor. The fall time of the voltage pulse is approximately 13 nanoseconds, and its half width is about 50 nanoseconds.

Except for the 9 KVDC max potential supplied to
the final stage amplifier, the various potentials required to
drive the above circuits are supplied by a single transistorized
power supply whose circuit is shown in Fig. 3.14. In order to
reduce the size of the power supply to a minimum, a 90 kHz
oscillator [22] , employing transistor Q23, is used. The os-

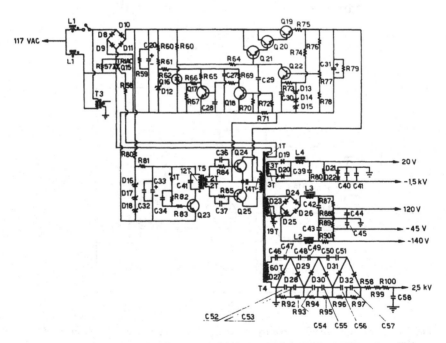

Fig. 3.14. Power supply circuitry for solid state laser modulator. Specifications of the
elementary components in this circuit are given in Tables 3.2 and 3.11 – 3.13.

cillator drives transistors Q24 and Q25 which chop the 110
VDC potential provided by the regulated power supply employing
transistors Q17-Q22 and silicon control rectifier Q16. The
resulting 200 V peak to peak rectangular wave is stepped-up
to 800 peak by transformer T4 to drive a Cockroft- Walton
voltage multiplier [22] yielding a 2.5 KVDC output. Other vol-
tages are obtained in a standard manner by the use of diode
circuits fed by various windings of transformer T4. A one-turn
winding on this transformer supplies the driving pulse to the
triac which, as previously described, is activated approximately 100 microseconds after the main switch is turned on, the
delay being provided by the RC circuit, controlled primarily
by resistor R65 and capacitor C33.

For proper control of the stroboscopic laser output, the amount of energy supplied by the pump lamps must be
carefully adjusted to a proper level that depends on the frequency of operation. For a given capacitor bank in the pump
lamp power supply, this is accomplished simply by regulating
its voltage output. The results are illustrated in Figs. 3.15a
and 3.15b (*) showing oscilloscope records of pulses sensed by a
high frequency light detector (EGG Lite-Mike) obtained at 3.3
KV and 3.1 KV, respectively. The first produces evidently too
strong excitation of the laser rod so that the intensity of
its light pulses is insufficiently controlled by the Kerr-cell

(*) See plates pp. 179 ff.

shutter, resulting in a series of uneven light bursts, unaccept-
able for uniform cinematography. The latter produces a series of
pulses whose amplitude is uniformly bounded within a Gaussian-
type envelope. Although the light intensity varies accordingly
throughout the operating period, the variation between successive
pulses is small and, with the added effect of a relatively low
sensitivity of photographic emulsion (Kodak High Speed infrared
film, 70 millimeters) one obtains a satisfactory uniform set of
cinematographic records, as those represented in Chapter 1.

            If the pump lamp power supply capacitor is charg-
ed to a potential which is too low, at first the triggering of
the pulse train becomes delayed, since it takes then a longer
time to build up the population inversion in the laser rod to
the threshold level. The delay is unreliable and the total number of pulses becomes, as the same time, decreased by various
amounts, until, at last, the laser discharges in form of a
single giant pulse. Below this limit one does not get, of course,
any laser output.

References

[1] McClung, F.J., and R.W. Hellwarth, "Giant Optical Pulsations from Ruby", <u>J. Appl. Phys.</u>, <u>33</u>, 828-829, 1962.

[2] Ellis, A.T. and M.E. Fourney, "Applications of a Ruby Laser to High-Speed Photography", <u>Proc. Inst. Electron. Engrs.</u>, <u>51</u>, 941-942, 1963.

[3] Schawlow, A.L. and C.H. Townes, "Infracted and Optical Masers", <u>Phys. Rev. 112</u>, 1940-1949, 1958.

[4] Maiman, T.H., "Stimulated Optical Emission in Fluorescent Solids", <u>Phys. Rev.</u>, <u>123</u>, 1145-1150, 1961.

[5] Hellwarth, R.W., "Control of Fluorescent Pulsations", <u>Advances in Quantum Electronics</u>, Columbia University Press, New-York, pp. 334-341, 1961.

[6] Collins, R.J. and Kisliuk, P;, "Control of Population Inversion in Pulsed Optical Masers by Feedback Modulation, <u>J. Appl. Phys.</u>, <u>33</u>, 2009-2011, 1962.

[7] Dunnington, F.G., "The Electrooptical Shutter – Its Theory and Technique", <u>Phys. Rev.</u>, <u>38</u>, 1506-1534, 1931.

[8] Jenkins, F.A. and White, H.E., <u>Fundamentals of Optics</u>, McGraw-Hill Book Company, Inc., New-York, 637 pp.

[9] Kingsbury, E.F., "The Kerr Electrostatic Effect", <u>Rev. Sci. Instrum.</u>, <u>1</u>, 22-32, 1930.

[10] Zarem, A.M., Marshall, F.R. and Hauser, S.M., "Millimicro second Kerr Cell Camera Shutter", <u>Rev. Sci. Instrum.</u>, <u>29</u>, 1041-1044, 1958.

[11] Pockels, F.C.A., <u>Einfluss des Elektrostatischen Feldes aut das Optische Verhalten Piezo-elektrischer</u>

Krystalle (Preisschrift) Akademie der Wissen-
schaften Gottingen, Abbandlungen der Gesellschaft,
39, 204 pp., 1894.

[12]   Pockels, F.C.A., Lehrbuch der Krystaloptik, B.G. Teubner,
Leipzig und Berlin, 1906.

[13]   Billings, B.H., "The Electro-Optic Effect in Uniaxial
Crystals of the Type XH PO , I. Theoretical",
J. Opt. Soc. Am., 39, 7 7- 01, 1949.

[14]   Sorokin, P.P., Luzzi, J.J., Lankard, J.R. and Pettit, G.D.,
"Ruby Laser Q-Switching Elements Using Phthalo-
cyanine Molecules in Solution", IBM Journal of
Res. and Dev., 8, 182-184, 1964.

[15]   Smith, W.V. and Sorokin, P.P., The Laser, McGraw-Hill
Book Co., Inc., New-York, 1966, 498 pp.

[16]   Soffer, B.H., "Giant Pulse Laser Operation by a Passive,
Reversibly Bleachable Absorber", J. Appl. Phys.,
35, 2551, 1964.

[17]   Kafalas, P., Masters, J.I. and Murray, E.M.E., "Photo-
sensitive Liquid Used as a Nondestructive Passive
Q-switch in a Ruby Laser", J. Appl. Phys., 35,
2349-2350, 1964.

[18]   Stark, T.E., Cross, L.A. and Hobart, J.L., "Saturable
Filter Investigation", a Technical Report pre-
pared for ONR, Feb. 19, 1964, under Contract
No. Nonr-4125 (00), NR015-702.

[19]   Cross, L.A. and Cheng, C.K., "Generation of Giant Pulses
from a Neodymium Laser with an Organic-Dye Saturable Filter", J. Appl. Phys., 38, 5, 2290-2294,
1967.

[20]   Lo, C.C., "A 2-M H_2 8-KV Pulser for High-Speed Stroboscopic Photography", Lawrence Radiation Laborato-
ry, UCRL-19248, University of California, Berkeley,
1969, 17 pp.

[21] Miller, Harold W. and Kerns, Q.A., "Transistors for
      Avalanche–Mode Operation", Rev. Sci. Intr., 33
      877–878, 1962.

[22] Lo, C.C., "1–kHz Spark Gap Trigger Amplifier", Lawrence
      Radiation Laboratory – Berkeley, Engineering Note
      EET–1303, March 1969.

TABLES FOR CHAPTER 3

TABLE 3.1

Kerr Constant for Different Substances (*)

Liquid	Formula	$BX10^{-7}$ (**)
Benzene	C_6H_6	0.6
Carbon Disulfide	CS_2	3.21
Chloroform	$CHC1_3$	-3.46
Water	H_2O	4.7
Nitrotoluene	$C_5H_7NO_2$	123.
Nitrobenzene	$C_6H_5NO_2$	220.

(*) Handbook of Physics, Condon and Odishaw,
    McGraw-Hill, 6-117.

(**) B in esu/cm or v/300 per cm.

TABLE 3.2

Chokes and Transformers Used in the
Solid State Laser Modulator

Item	Core Type	Winding Description
T1	Indiana General CF101, Q-2 toroid	Primary 4 turns evenly distributed. Secondary: 3 turns on one half of the core, another 3 turns on the other half of the core.
T2	Indiana General CF122, H toroid (2 toroids)	Primary: 10 turns evenly distributed. Secondary 4 turns on one half of the core, another 4 turns on the other half, connected in parallel.
T3	Triad F16X	
T4	Indiana General CF123 0-5 toroid (2 toroids)	Primary: 14 turns twisted pair evenly distrubuted connected in series. Secondary: 19 turns twisted pair evenly distributed, connected in series; 25 turns twisted pair evenly distributed connected in series.
T5	Ferroxcube 846T250 3E2A toroid	Primary: 12 turns evenly distributed. Secondary turns C.T.
L1	Indiana General CF123, 0-5 toroid (2 toroids)	15 turns twisted pair evenly distributed. # 16 plastic insulated.
L2	Ferroxcube	20 turns evenly distributed;
L3	846T250	
L4	3E2A toroid	NOTE: All Wire # 20 AWG, Teflon insulated. Round off edges of all toroids.

TABLE 3.3

Resistors in Fig. (3.10)

Resistor	Resistance		Power
R1	560	Ω	1/2 w
R2	16	kΩ	1/2 w
R3	11	kΩ	1/2 w
R4 – R5	1.2	kΩ	1/2 w
R6	2	kΩ	1/2 w
R7 – R8	62	kΩ	2 w
R9	120	kΩ	1/2 w
R10	47	kΩ	1/2 w

TABLE 3.4

Capacitors in Fig. (3.10)

Capacitor	Capacitance	Type
C1	220 pf	Disc
C2	1,100 pf	Disc
C3	220 pf	Disc
C4	22 μf	Electrolitic, 35V
C5	0.1 μf	Disc

TABLE 3.5

Transistors and Diodes in Fig. (3.10)

Item	Specification
Q1 – Q3	2N3643
Q4	2N2219
D1 – D2	1N4447

TABLE 3.6

Resistors in Figs. (3.11) and (3.12)

Resistor	Resistance	Power	Resistor	Resistance	Power
R11	51 Ω	1/2 w	R25	1 kΩ	1/2 w
R12	200 Ω	1/2 w	R26	2 kΩ	1/2 w
R13	2 kΩ	1/2 w	R27	1 kΩ	1/2 w
R14	51 Ω	2 w	R28	300 kΩ	1/2 w
R15	1.2 kΩ	1/2 w	R29 – R30	1 kΩ	1/2 w
R16	1 kΩ	1/2 w	R31	100 kΩ 10 turns	1/2 w
R17	2 kΩ	1/2 w	R32	9.1 kΩ	1/2 w
R18 – R19	1 kΩ	1/2 w	R33 – R34	1 kΩ	1/2 w
R20	50 kΩ	1/2 w	R35	9.1 kΩ	1/2 w
R21	3.9 kΩ	1/2 w	R36	1 kΩ	1/2 w
R22 – R23	1 kΩ	1/2 w	R37	470 Ω	1/2 w
R24	9.1 kΩ	1/2 w	R38 – R39	20 Ω	1/2 w
			R40	5.1 Ω	1/2 w

TABLE 3.7

Capacitors in Figs. (3.11) and (3.12)

Capacitor	Capacitance	Type
C6	0.001 μf	Disc
C7	0.1 μf	Disc
C8	0.002 μf	Disc
C9 – C10	0.001 μf	Disc
C11	0.013 μf	Disc
C12	0.1 μf	Disc
C13	22 μf	Electrolitic, 35V
C14	0.1 μf	Disc
C15	180 μf	Disc
C16	0.1 μf	Disc

TABLE 3.8

Transistors and Diodes in Figs. (3.11) and (3.12)

Item	Specifications
Q5 – Q12	2N2219
Q13 – Q14	2N2222
D3 – D6	1N4447

TABLE 3.9

Resistors in Fig. (3.13)

Resistor	Resistance	Power
R41	11 kΩ	1 w
R42 – R44	360 Ω	1 w
R45 – R48	1.2 kΩ	2 w
R49 – R56	2.7 Ω	1 w

TABLE 3.10

Capacitors in Fig. (3.13)

Capacitor	Capacitance	Type
C17 – C19	0.004 μf	Disc; 4kV
C20 – C22	0.002 μf	Disc
C23	0.1 μf	Disc
C24	1 μf	Mylar
C25	0.04 μf	Disc; 10 kV

TABLE 3.11

Resistors in Fig. (3.14)

Resistor	Resistance	Power	Resistor	Resistance	Power
R57	20 Ω	1 w	R76	56 kΩ	1/2 w
R58	200 Ω	1 w	R77	10 kΩ	1/2 w
R59	20 kΩ	2 w	R78	27 kΩ	1/2 w
R60 – R61	7.5 kΩ	2 w	R79	20 kΩ	2 w
R62	10 Ω	1/2 w	R80	2 kΩ	15 w
R63	18 kΩ	2 w	R81	10 Ω	1/2 w
R64	3.3 kΩ	2 w	R82	330 kΩ	1/2 w
R65	5.1 kΩ	1/2 w	R83	7.5 kΩ	1/2 w
R66	10 kΩ	1/2 w	R84 – R85	100 Ω	1/2 w
R67	1 kΩ	1/2 w	R86	47 kΩ	1 w
R68	10 kΩ	1/2 w	R87	4 kΩ	2 w
R69	2.2 kΩ	1/2 w	R88	18 kΩ	2 w
R70	2.4 kΩ	1/2 w	R89	6.8 kΩ	2 w
R71 – R72	1 Ω	1/2 w	R90	13 kΩ	2 w
R73	10 Ω	1/2 w	R91	2 kΩ	2 w
R74	82 kΩ	1/2 w	R92 – R97	22 MΩ	1/2 w
R75	1 Ω	1 w	R98 – R100	1.6 kΩ	1/2 w

TABLE 3.12

Capacitors in Fig. (3.14)

Capacitor	Capacitance		Type
C26	1,000	f	Electrolitic; 450 V
C27 – C28	0.1	f	Disc
C29	0.33	f	Disc
C30	0.1	f	Disc
C31	16	f	Electrolitic; 450 V
C32	0.1	f	Electrolitic; 75 V
C33	1,000	f	Electrolitic; 50 V
C34	250	f	Disc
C35	5,100	f	Disc
C36 – C37	0.003	f	Disc
C38	1	f	Mylar; 200 V
C39	2	f	Mylar; 100 V
C40	0.1	f	Disc
C41	22	f	Electrolitic
C42 – C45	0.47	f	Disc
C46 – C47	0.1	f	Disc; 600 V
C58	1	f	Mylar; 3 kV

TABLE 3.13

Transistors and Diodes in Fig. (3.14)

Item	Type	Item	Type
Q15	GESC45D (Triac)	D8 – D11	1N3775
Q16	2N2326	D12	1N3029B
Q17	2N2209	D13 – D15	1N942A
Q18	2N2219	D16 – D18	1N3022B
Q19 – Q20	2N3902	D19 – D20	UTR60
Q21	2N5059	D21 – D22	UT71
Q22	2N5058	D23 – D26	UTR60
Q23	2N3019	D27 – D32	USR15
Q24 – Q25	2N3902		

PLATES FOR CHAPTER 3

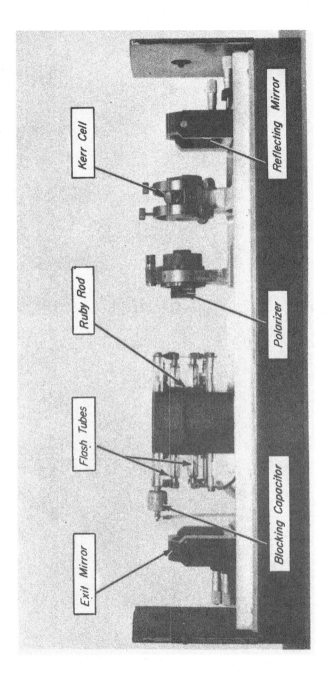

Fig. 3.4. Photograph of straight laser head.

Fig. 3.5. Photograph of Brewster-angle laser head.

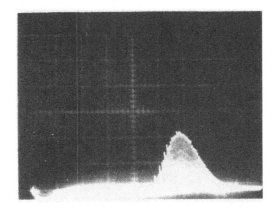

—a—

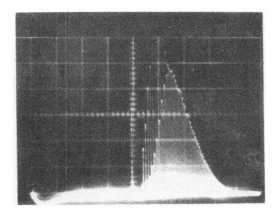

—b—

Fig. 3.15. Oscilloscope records of light output of Kerr cell modulated ruby laser. Modulation frequency is 200 kHz; sweep speed is 100 micro-seconds per cm. The pump lamp voltage is :

  a) 3.1 kV, and
  b) 3.3 kV.

APPENDIX A

BIREFRINGENCE AND POLARIZATION

A.1 Introduction

Since light polarization techniques play a fundamental role in laser cinematography, physical principles of the operation of devices used for this purpose are briefly reviewed. The information presented here is based primarily on the texts of Jenkins and White [1] (*) and Born and Wolf [2] to which the reader interested in more details is referred.

Described here first is the phenomenon of double refraction whose knowledge dates from the studies of Huygens who in 1678 discovered that, if a calcite crystal is exposed to a ray of unpolarized light, two refracted plane polarized rays are obtained. The fundamental principles of this phenomenon are then discussed with particular reference to calcite crystal as the simplest device possessing such a property. A similar property exhibited by quartz which is used as an essential element of the neutral wedge in the laser schlieren system described in Chapter 2, differs from that of calcite only as a consequence

_____

(*) Numbers in square brackets denote references listed at the end of Appendix A.

of a more complicated geometry of the crystal lattice. This,
however, is for the present purpose unessential, especially
since the pertinent information can be easily found in the reference text [1] .

A.2 Birefrigerent Crystals

The best known crystals possessing the property
of double refraction or birefringence, are calcite (i.e. calcium
carbonate, $CaCO_3$) or quartz (i.e. silicon oxide, SiO_2). These
crystals are anisotropic and uni – axial in structure that is,
whereas their physical properties are, in general, dependent on
direction, they possess a line of symmetry with respect to both
the geometry of the crystal and the arrangement of its atoms,
known as the optic axis. In the direction perpendicular to this
axis physical properties of the crystal are the same, while at
any other angle properties change, reaching a maximum or a minimum along the axis.

A.3 Polarization by Birefrigerence

When an unpolarized light ray is incident upon
a double refracting crystal at an angle to the optic axis, two
refracted rays emerge. The geometry of this phenomenon as it
occurs in a calcite crystal is shown in Fig. A.1. Only one of

the rays, how-
ever, obeys the
Snell law of re
fraction, accord
ing to which

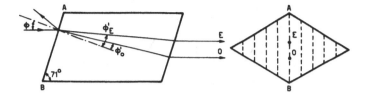

Fig. A.1. Double refraction of light in a calcite crystal.

$$n = \frac{\sin \Phi}{\sin \Phi'} \qquad (A.1)$$

where n is the relative refraction index of the crystal with
respect to air, while Φ and Φ' are the angles of incidence and
refraction, respectively. This ray is, therefore, called ordi-
nary or O-ray, while the other is referred to as the extraor-
dinary or E-ray.

Since the opposite sides of calcite crystals are
parallel, the two refracted rays emerge parallel to each other,
as well as to the incident ray. Only the O-ray, however, remains in the plane of incident and reflected rays.

If an unpolarized light ray is incident upon the
crystal in the direction of the optic axis, the phenomenon of
double refraction does not occur.

A.4. Rotation of Polarization by Birefringence

If the front face of a polarizing crystal, that
is, a face parallel to its optic axis, is exposed to a plane polarized light ray incident upon it at a right angle, while the

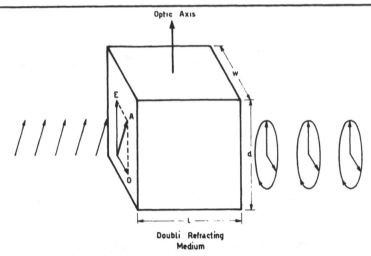

Fig. A.2. Production of elliptically polarized light by the use of a doubly refracting medium.

plane of polarization is at a non-zero angle to the optic axis, as illustrated by Fig. A.2, the light vec̲tor $\underline{A}$ is divided into an E-ray, parallel to the optic axis, and an O-ray, perpendicular to it. Due to the anisotropy of the crystal, the two rays experience two different refractive indices, n_E and n_0 respectively, yielding a difference in their optical path lengths,

$$(A.2) \qquad \Delta P = \frac{1}{\lambda_v}(n_E - n_0)l$$

where l is the length of the crystal and λ_v is the light wavelength in vacuum.

On emerging from the crystal, the difference in phase δ between the two components is

$$(A.3) \qquad \delta = \frac{2\pi}{\lambda_v}l(n_E - n_0) \ .$$

Thus, as the phase difference between the two components is incresed by increasing the optical path difference ΔP, the emerging beam continuously changes its degree of elliptical polarization. When the angle between the plane of polarization of the incident ray and the optic axis is 45°, the two vector components of the E and O-ray are equal. The ellipticity of polarization resulting from such a geometry is a function of the phase angle δ, as shown in Fig. A.3. As it is clear from Eq. (A.3), for a given light frequency, this can be controlled simply by varying the length l.

A.5 Polarizing Elements

The phenomenon of double refraction can be used to produce polarized light in a number of devices. Of these, the most relevant to our purpose is the polarizing prism [1]. Such a prism can be made out of two halves of a calcite crystal which is appropiately cut and cemented together in such a manner that one ot the two refracted beams is removed by total reflection, as shown in Fig. A.4, producing thus a plane polarized light ray. The angles shown in Fig. A.4 are for the Nicol prism using Ca—

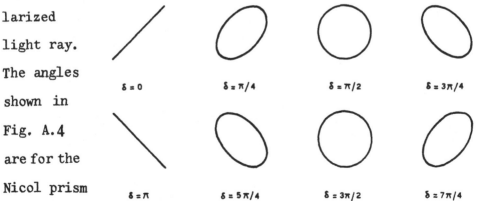

Fig. A.3. Change of the ellipticity of polarized light as a function of the phase difference, δ.

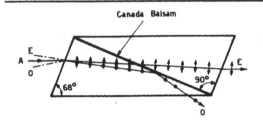

Fig. A.4. Production of plane polarized light by the use of Nicol prism.

nada balsam as the cementing agent. When the end faces of the prism are cut at right angles to the sides, so that the light enters and leaves in the direction normal to the front surface, the prism is referred to as a Glan-Thompson polarizer. If, in the geometry of Fig. A.4, an air gap is used instead of the Canada balsam, it is called a Foucault polarizer.

References

[1] Jenkins, F.A. and White, H.E., <u>Fundamentals of Optics</u>, McGraw-Hill Book Company, Inc., New-York, 1957, 637pp.

[2] Born, M. and Wolf, E., <u>Principles of Optics, Electromagnetic Theory of Propagation, Interference and Diffraction of Light</u>, Pergamon Press, New-York, 1959, 803 pp.

APPENDIX B

CONTINUOUS FLOW MIXING OF EXPLOSIVE GASES

B.1 Introduction

The usual means of preparing an explosive mixture of
gases is by batch mixing, whereby each constituent is introduced into a chamber of some fixed volume and its desired a-
mount metered by partial pressure. The final homogeneous state
is reached by convective and molecular diffusion after a suitable time interval. This method restricts the experimenter to
a single composition per filling, and it is associated with
many inconveniences, especially the unavoidable time delay and
the ever-present danger of explosion.

The continuous flow mixing apparatus combines
versatility with safety and ease of operation. The flow rate of
each constituent is monitored solely by the upstream stagnation
pressure so that any desired change in composition of the mixture can be obtained by just the turning of a knob, and the
homogeneous state is attained by turbulent flow mixing in the
outlet manifold whose extremely small internal volume provides
ample safeguard against explosion.

Operation of the apparatus is based on the use of

commercially available hypodermic needles and small diameter
tubes as the primary metering elements. The reliability is as-
sured by three conditions:

    (i) maintenance of chocked flow at tube exit,

    (ii) restriction of the friction coefficient within the
         range of laminar Reynolds numbers and

   (iii) operation at sufficiently high density so that the sub
         stance behaves as a continuum.

As a consequence of the first, the flow is monitored by a single para
meter, the upstream stagnation pressure, and the flow through the
metering tube is indipendent of any downstream disturbances. The
second and third conditions impose, respectively, the upper and the
lower bounds on the operating range for a given metering tube, and
they necessitate the use of a proper set of such tubes to cover a
sufficiently wide range of compositions. For this reason provison
has been made in the construction for a simple and reliable ex
change of metering tubes. The syringe fitting for hypodermic
needles has been found especially convenient in this respect.

B.2 Construction

      Photographs of the apparatus are shown in Fig.
B.1 (*); its flow diagram is given by Fig. B.2. Each gas is led
from the commercial supply cilinder S , provided with a re-

(*) See plates pp. 213 ff.

ducing valve, to a Grove
(Grove Valve and Regula-
tion Co.) regulator RV
Pressure in the line is
measured by a 0-30 lb.
in^{-2} precision gauge P,
and the gas is admitted
through a Skinner (Skinner
Chuck Co.) two way nor-
mally closed explosion-
proof solenoid valve,

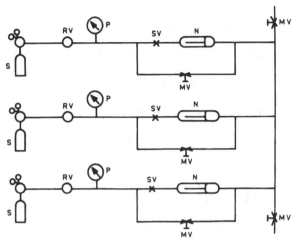

MV Manual Valve   P Pressure Gauge   S Gas Supply
N Needle Holder   RV Regulation Valve   SV Solenoid Valve

Fig. B. 2. Schematic flow diagram.
MV, manual value; N, needle; P, pressure gauge; RV, regulating valve; S, gas supply ; SV, solenoid valve.

type X52, SV to the needle holder assembly N , whose construction is described by Fig. B.3. In order to facilitate bleed-
ing, each line is provided with a by-pass of the solenoid valve

NEEDLE-HOLDER   ASSEMBLY

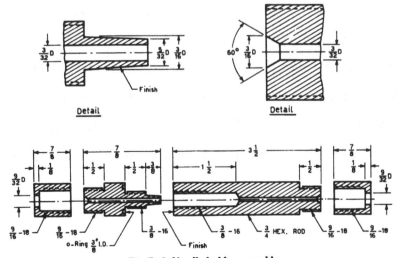

Fig. B. 3. Needle-holder assembly.
All dimensions in inches.

and needle-holder assembly fitted with a high pressure needle valve MV . After passing the metering tube, the gas enters the mixing manifold from which it is admitted to the experiment al apparatus through one of the two high-pressure needle type exit valves MV . All lines are made out of 1/16 inch i.d. Autoclave (Autoclave Engineers) stainless-steel tubing. For mobility the whole apparatus, including the gas supply cylinders, is mounted on a four-wheeled structure made out of Unistrut (Unistrut Corp.) elements.

The metering tubes are - for most part - the Yale (Beckton, Dickson and Co.) Regular Point, hyperchrome stainless hypodermic needles. These needles are usually avail- able only to gauge 30(D = 0.006 in). In order to attain small er sizes, commercially available tubes of small internal dia- meter have been inserted into larger needles and secured in place by gliptal paint. The tubes have been cut, ground and honed to the required length with all the burr, caused by this process, carefully removed under a microscope. All the needles are mounted facing the flow in order to provide an unobstructed Borda mouthpiece entry and allow the pressure upstream to push the needle against the tapered end of the holder and thus pre- vent leakage.

B.3 Operation

(a) Flow rate

Under proper operating conditions, the flow through the metering tubes is controlled by the chocked Fanno process [1] (*). For a perfect gas with constant specific heats this implies that the molecular flow rate is given by:

$$\dot{N} = \frac{\pi D^2 \dot{m}}{4 W} = \frac{\pi D^2}{4 W} = \frac{\gamma P_o M}{a_o}\left(1 + \frac{\gamma - 1}{2}M^2\right)^{-\frac{\gamma + 1}{2(\gamma - 1)}} \qquad (B.1)$$

where p_o and a_o are the stagnation pressure and speed of sound respectively, γ is the specific heat ratio of the gas, W its molecular weight, while D is the inside diameter of the tube. The Mach number, M, at the tube inlet appears in the above as a parameter which, in turn, is related to the friction coefficient f as follows:

$$\frac{4fL}{D} = \frac{1 - M^2}{\gamma M^2} + \frac{\gamma + 1}{2\gamma}\ln\left\{\frac{(\gamma + 1)M^2}{2\left[1 + \frac{\gamma - 1}{2}M^2\right]}\right\} . \qquad (B.2)$$

If the flow is laminar

$$f = \frac{16}{R_e} = \frac{4\pi \mu D}{\dot{N} W} \qquad (B.3)$$

where L is the tube length, and μ is the gas viscosity coef-

(*) Numbers in square brackets denote references listed at the end of Appendix B.

ficient.

Equations (B.2) and (B.3) yield the second parametric expression for the flow rate, namely:

(B.4) $\dot{N} = \dfrac{16 \pi \mu L D}{W} \left\{ \dfrac{1 - M^2}{\gamma M^2} + \dfrac{\gamma + 1}{2\gamma} \ln \dfrac{(\gamma + 1)M^2}{2 \left[1 + \dfrac{\gamma - 1}{2} M^2 \right]} \right\}$.

If we know the speed of sound for each gas at room temperature a_o , Eqs. (B.1), (B.3) and (B.4) can be solved for a particular tube geometry to find the molecular flow rate, the Reynolds number and the Mach number in terms of a single parameter: the stagnation pressure p_o .

Figures B.4 – B.9 (*) present the results for some typical gases used in the research on gasdynamics of explosions, namely: hydrogen, helium, acetylene, nitrogen, oxygen and argon at room temperature ($T = 70°F$) as a function of the stagnation pressure for tube diameters in the range from 0.0046 – 0.0075 in.

(b) Critical flow condition

In order to assure chocked flow through the metering tube, the pressure downstream must not exceed the critical value specified by:

(B.5) $p^* = M \left(\dfrac{2}{\gamma + 1} \right)^{1/2} \left[1 + \dfrac{\gamma - 1}{2} M^2 \right]^{-\frac{\gamma + 1}{2(\gamma - 1)}}$

(c) Laminar flow condition

For reliability of operation it is most important to avoid the transition regime, since there friction coef-

(*) See plates pp. 213 ff.

ficient can have different values depending on starting and local flow conditions. This imposes the following upper bound on the flow rate:

(B.6)
$$\dot{N}_{Lim} = \frac{\pi\, D \mu}{4W}\, Re_{Lim} \; .$$

The effect of this bound is manifested in Fig. B.6 by the departure of experimental points from theoretical curves at high values of $\dot{N}$, in contrast to the excellent a-greement at lower Reynolds numbers. This is confirmed by Fig. B.8 which displays the experimental results obtained with helium corresponding to particularly low Reynolds numbers.

Taking $Re_{Lim} = 1750$, values of stagnation pressures giving maximum flow rates in accordance with this condition, for a variety of gases, are presented on Fig. B.7 as a function of the tube diameter.

(d) Continuum flow condition

In order to make sure that the gas medium behaves as a continuum in the course of flow through small diameter tubes, the numbers n of molecules per unit volume cannot be less than:

$$n_{Lim} = \frac{1}{\sqrt{2}\,\pi\,\lambda_{Lim}\sigma^2}$$

where λ is the molecular mean free path and σ the molecular diameter.

By postulating that the limiting mean free path

λ_{lim} should not exceed 0.01 D and using perfect gas equation of state, this leads to the following condition for operating pressure below which slip flow regime may be set in the tube

$$p_{lim} = 0.998 \times 10^{-4} \frac{T}{D G^2} \tag{B.8}$$

where p_{lim} is in lb in^{-2} abs., D in inches, G in Å and T in °R . With the use of the Loschmidt formula [2] :

$$G^2 = \frac{10^{-24}}{78.26 \mu} (3RTW)^{1/2} \tag{B.9}$$

this becomes:

$$p_{lim} = 78.26 \times 10^{20} \frac{\mu}{D} \left(\frac{T}{3RW}\right)^{1/2} \tag{B.10}$$

where μ is in lb. in.$^{-1}$s.$^{-1}$, R in ft. lb. mole^{-1} degr^{-1} and W in lb. mole^{-1} . Since p_{lim} is equivalent to p* in Eq. (B.5), the limiting value of stagnation pressure $p_{o\,lim}$, below which slip flow may occur inside the tube is given by:

$$p_{o\,lim} = \frac{78.26 \times 10^{20} \mu}{D M} \left(\frac{T}{3RW} \frac{\gamma+1}{2}\right)^{1/2} \left[1 + \frac{\gamma-1}{2} M^2\right]^{\frac{\gamma+1}{2(\gamma-1)}} . \tag{B.11}$$

Of all the gases considered, this limit has been found to affect only hydrogen and helium. Figure B.10 presents the limiting stagnation pressure for these two gases as a function of tube diameter. Figure B.11, however, demonstrates that the effect of slip flow can be neglected for our pressure range, 0 – 30 lb. in.$^{-2}$ gauge, even below the lower pressure limit of

Eq. (B.11)

(e) Metering tube condition

Besides the fluid-dynamic considerations, the operation of the metering tube may be affected by heat transfer and geometry.

According to the theory of the Fanno flow, the process should be adiabatic. In order to test this condition, a metering tube has been coated with an insulating paint and its performance recorded before and after the insulation, Fig. B.7. The insulated tube has been found to behave more in accord with the theory, but the improvement has not been significant.

As far as geometry is concerned, one has to take under consideration the dimensional tolerance and the inner surface condition of commercially available tubes, as well as the shape of their tips. Even though reported tolerances are as high as 30% [3], their practical effect is offset by the corrugations of the inner surface thereby reducing the scatter in results. Using a variety of needles whose size was chosen at random and calibrated under the same operating conditions, we have found that deviation in flow rates does not exceed 6%. Conservatively then, if one is not interested in an absolute accuracy higher than 10%, theoretical curves are sufficiently reliable. For higher accuracy each needle must be individually calibrated.

The tip of a hypodermic needle is slanted. This

upsets the symmetry of flow at the inlet and consequently in-
droduces the transition regime at a lower Reynolds number than
predicted. Moreover a nominal 1 in. needle has an actual length
of about 1.2 in. from the tip to the end inside the holder. In
order to check the effect of these conditions on the performance, tests were made with needles cut to 1 in. long with the
tip ground flat and honed. A sample of results is shown in Fig.
B.7 indicating a marked improvement, exceeding in fact that
accomplished by insulation. Particular care has to be exercised
in keeping the tip free of any distortion associated with the
presence of dirt or burr, since deviation as large as 50% have
been recorded when cleanliness of the tip has not been ascer-
tained by microscopic inspection.

(f) Calibration

Each tube has been calibrated by the use of a
five-gallon bottle provided with an accurate pressure gauge.
For each stagnation pressure the flow rate has been determined
by the use of a perfect gas equation of state according to
which

(B.12)
$$\dot{N} = \frac{V}{RT}\frac{dp}{dT} \;.$$

A representative set of results for a hypodermic
needle, gauge 30, with nitrogen as the flowing substance, has
been shown on Fig. B.6.

B.4 Sample Problem

In order to illustrate the use of the theoretic-
al flow rate curves presented in Figs. B.4 to B.9 and the above
mentioned calibration technique. Let us consider a typical laboratory problem that is specified as follows: It is required
to fill a 0.157 ft^3 shock tube in a period of 60 sec. with the
following mixture:

$$1O_2 + 2H_2 + 5N_2$$

where the final pressure in the tube ranges from 100 to 200
mmHg at room temperature.

The first step in the solution is to find the
partial pressures of every gas species, p_x , corresponding
to $p_t = 100$ mmHg and to $p_t = 200$ mmHg, where p_t is the
total pressure for all species. The required minimum and maximum flow rates for each species, are given by Eq. (B.12), expressed simply as

$$\dot{N}_x = \frac{V}{RT} \frac{p_x}{t} \qquad (B.13)$$

since, as verified experimentally, p_x varies linearly with
Substituting for $V = 0.157$ ft^3 and $t = 60$ sec. Eq. (B.13) yields:

$$\dot{N}_x = 8.75 \times 10^{-9} p_x \qquad (B.14)$$

where p_x is taken in mmHg, and $\dot{N}_x$ in lb. -mole/sec. The results for p_x and $\dot{N}_x$ are presented in Table B.1.

From Fig. B.8 for oxygen it is found that the range of required flow rates is covered by a tube diameter of 0.006", i.e. gauge number 30. Figure B.4 shows that the required flow rate range for hydrogen can be covered either by a tube diameter of 0.005" alone or by both the diameters of 0.004" and 0.006", where $L = 1"$ for both oxygen and hydrogen.

In the case of nitrogen, however, the large values of the required flow rates cause the flow to be non-laminar and not amenable to the theoretical treatment of this paper. Hence a relatively large diameter tube should be calibrated for nitrogen, according to section B.3(f), in order to obtain its $\dot{N}$ vs. p_0 curve. Figures B.12 and B.13 present the calibration curves for a one inch, gauge 26 hypodermic needle.

References

[1] Keenan, J. and Kaye, J., <u>Gas Tables</u>, John Wiley, New York, 1948, pp. 209–211.

[2] Present, R.D., <u>Kinetic Theory of Gases</u>, McGraw-Hill, New-York, 1958.

[3] <u>Superior Tube Data Memorandum N°9</u>, Superior Tube Co., Norristown, Pa.

TABLES FOR APPENDIX B

TABLE B.1

Results of Sample Problem

Species	Px (min) mm—Hg	Px (max) mm—Hg	N_x (min) lb-mole/sec	N_x (max) lb-mole/sec
O_2	12.5	25.0	1.09×10^{-7}	2.18×10^{-7}
H_2	25.0	50.0	2.18×10^{-7}	4.36×10^{-7}
N_2	62.5	125.0	5.42×10^{-7}	10.84×10^{-7}

PLATES FOR APPENDIX B

Fig. B.1. Photographs of the apparatus: (a) front and back views, total dimensions 2 ft. x 2 ft. x 5–1/2 ft. ; (b) needle-holders mounted on an aluminum 3 in. x 17 in. plate.

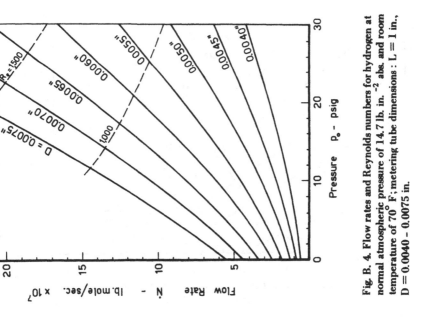

Fig. B. 5. Flow rates and Reynolds numbers for helium at room temperature; metering tube dimensions: L = 1 in., D = 0.0035 - 0.0080 in.

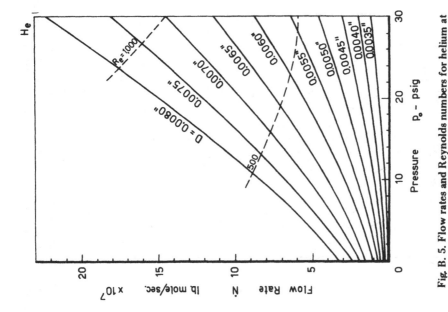

Fig. B. 4. Flow rates and Reynolds numbers for hydrogen at normal atmospheric pressure of 14.7 lb. in.$^{-2}$ abs. and room temperature of 70° F; metering tube dimensions : L = 1 in., D = 0.0040 - 0.0075 in.

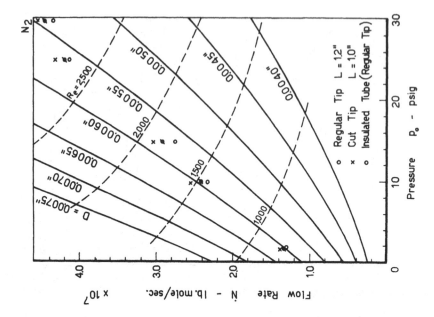

Fig. B. 7. Flow rates and Reynolds numbers for nitrogen at room temperature including the demonstration of the effect of the tip shape and insulation for a needle gauge 30 ; metering tube dimensions : L = 1 in., D = 0.0060 – 0.0075 in.

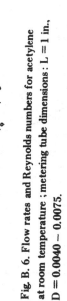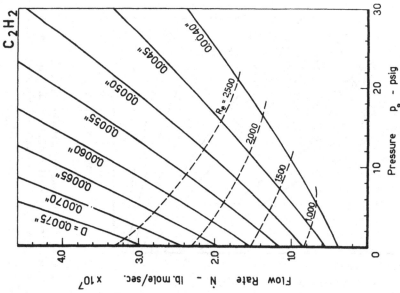

Fig. B. 6. Flow rates and Reynolds numbers for acetylene at room temperature ; metering tube dimensions : L = 1 in., D = 0.0040 – 0.0075.

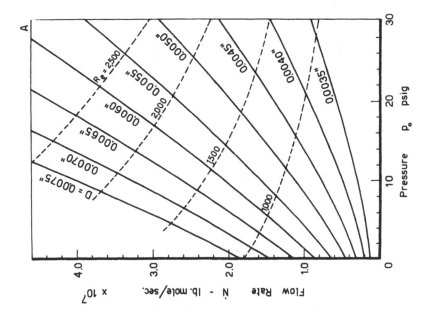

Fig. B. 9. Flow rates and Reynolds numbers for argon at room temperature; metering tube dimensions : L = 1 in., D = 0.0035—0.0075 in.

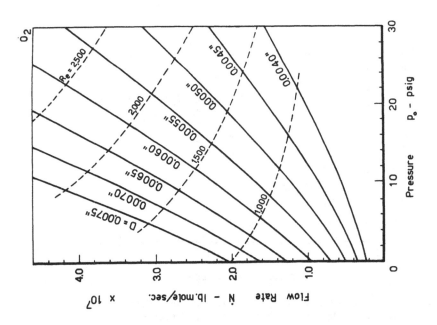

Fig. B. 8. Flow rates and Reynolds numbers for oxygen at room temperature; metering tube dimensions : L = 1 in., D = 0.0040—0.0075 in.

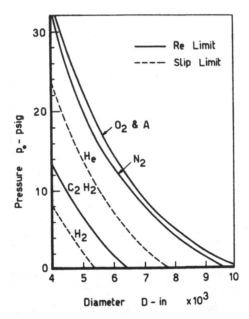

Fig. B. 10. Limiting stagnation pressure for
continuum and laminar flow conditions at
room temperature.

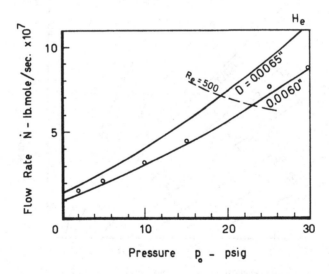

Fig. B. 11. Experimental flow rates for hypodermic needle gauge 30 using helium as the flowing substance;
solid lines represent theoretical results for L = 1 in., D = 0.0060 and 0.0065 in.

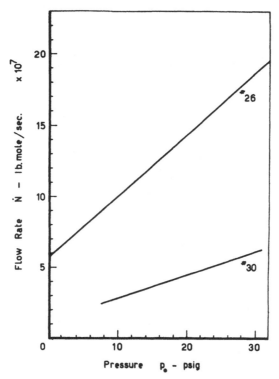

Fig. B. 12. Rate of Pressure increase, for 1" hypodermic needle gauge no. 26, as a function of the stagnation pressure at room temperature for nitrogen needle calibration of the sample problem. V   0.717 ft.

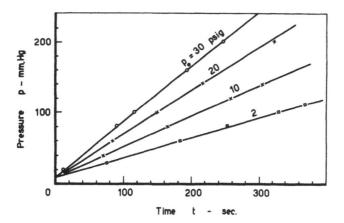

Fig. B. 13. Nitrogen needle calibration curve in the sample problem. Gauge no. 26 at room temperature. Calibration curve for gauge no. 30 is included for comparison.

Appendix C. Nomenclature

l — length of laser material

   — length of EOLM

M — Mach number

   — magnification

$\dot{m}$ — Mass flow rate in Appendix B

N — number density of atoms

$\dot{N}$ — molecular flow rate in Appendix B

n — extent of overlap

   — refractive index

o — object size (or distance)

P — optical path

p — pressure

   — image position

Q — optical path component in variable shear optics

R — perfect gas constant

Re — Reynolds number

r — distance or radius

s — shear

T — absolute temperature

t — time

V — voltage

   — volume in Appendix B

W – distance defined in Fig. 2.8

   – molecular weight in Appendix B

w – width of electro-optical light modulator EOLM

   – blast wave velocity

X – depth of test section

x – space coordinate r/r_n

y – a_a^2 / w_n^2

   – fringe spacing

   – lateral position in the test section

a – speed of sound

   – plate thickness

   – distance between grating and real image of test object in variable shear optics

B – Kerr constant

C – capacitance

D – distance defined in Fig. 28

   – inside diameter of metering tube in Appendix B

d – beam diameter

   – height of electro-optical modulator (EOLM)

E_0 – energy of formation for blast wave

E_{01} – energy difference between levels 0 and 1

E – amplitude of electric field vector

F – focal length of schlieren mirror

f – distance defined in Fig. 2.6

   – friction coefficient in Appendix B

G_A – absorption

G_E – amplification

G_R – attenuation

G – net gain

h – distance defined in Fig. 2.6

I – illumination

i – image size (or distance)

– driving current

j – geometric factor = 0, 1 and 2 for plane, cylindrical and spherical symmetries, respectively.

k – Boltzmann's constant

L – metering tube length in Appendix B

l – distance defined in Fig. 2.6

– line spacing in a grating

α – incidence angle

– absorption coefficient

γ – specific heat ratio

δ – fringe spacing

– phase shift between polarized light components

θ – light deflection angle

κ – dielectric constant

λ – wave-length

μ – front velocity modulus for blast waves

– viscosity coefficient in Appendix B

σ – absorption cross-section

$\quad$ – molecular diameter in Appendix B

Φ – angle of deflection

$\quad$ – incidence angle

$\quad$ – refraction angle

ψ – diffraction angle

SUBSCRIPTS

0 – ground level

1 – excited level

a – ambient condition

n – condition behind blast wave

r – reference exposure

t – text exposure

v – vacuum

Printed in the United States
By Bookmasters